Natural
Parasite Control
for Livestock

Pasture Management, Chemical-Free Deworming, Growing Antiparasitics, and More

WENDY LOMBARDI

Skyhorse Publishing

Skyhorse Publishing books may be purchased in bulk at special discounts for sales promotion, corporate gifts, fund-raising, or educational purposes. Special editions can also be created to specifications. For details, contact the Special Sales Department, Skyhorse Publishing, 307 West 36th Street, 11th Floor, New York, NY 10018 or info@skyhorsepublishing.com.

Skyhorse® and Skyhorse Publishing® are registered trademarks of Skyhorse Publishing, Inc.®, a Delaware corporation.

Visit our website at www.skyhorsepublishing.com.

10 9 8 7 6 5 4

Library of Congress Cataloging-in-Publication Data is available on file.

Cover design by Mona Lin
Cover photos by Getty Images

Print ISBN: 978-1-5107-5710-3
Ebook ISBN: 978-1-5107-5937-4

Printed in China

The information here within is strictly the opinion of the author. It is *not* intended to be a veterinary supplement or reference for treatment of livestock. The author is not held responsible for outcomes to livestock if individuals choose to implement the information within these pages. It is the responsibility of individuals to do further research to determine if the described treatments are adequate for their own needs.

We are the caretakers of any living creature which we confine, manipulate, or utilize. Therefore, it's our responsibility to give the best possible care to those creatures.

I am grateful to every beautiful life that shared my journey, taught me so much, gave so much.

Contents

Introduction vii

Internal Parasites, Hosts, and Life Cycles 1

External Parasites 11

Pasture Management 23

Eliminating Parasites with Antiparasitics 34

    List of Antiparasitics 39

Growing, Collecting, Processing, and/or Storing Antiparasitics 54

Performing At-Home Fecal Sample Checks 73

Final Thoughts 76

Resources 80

References 81

Index 88

Introduction:
It Doesn't Have to Be a Full-Time Job

During the course of your regular farm day, you might notice your chickens' combs just aren't as bright red as you've seen in the past. Maybe a cow or horse appears a little "ribby." Or, you look at your sheep's eyelids and they're not as bright pink as they should be. Perhaps there are a couple animals that never obtain the luster others possess in the same group. These are just a few signs parasites are "sucking the life" out of your livestock. As diligent as you may be, keeping parasites at bay may feel like a relentless, full-time job—especially if you want to do it without the use of chemicals.

But, by developing a system that easily works for you and your livestock, your job of parasite control can become a part-time, seasonal position! This can be accomplished 1) through good pasture management and rotation, 2) with environmental control of parasites, 3) by targeted deworming with natural products (possibly even growing some ingredients!), and, 4) with continuing research of parasite control in livestock.

There are specific symptoms for some parasites. Chart A (page 6) lists those indicators. However, this list is *not* a definitive diagnosis because a high load of any parasitic infestation can result in diarrhea, dehydration, weight loss, un-thriftiness, and/or lethargy no matter which parasite is doing the damage. If after reading and implementing the information in this book you still seem to have a parasite issue, you may want to collect a fecal sample to take to a

A pale comb and waddles is an indication of possible parasites.

Healthy chickens should have bright combs and shiny feathers.

qualified vet or lab to have the parasite correctly identified. This way, you can target treatment for that specific parasite. (Collecting and checking samples at home will be discussed later.)

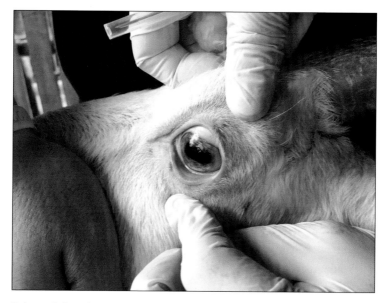

Pale eyelids indicate anemia, usually caused by a heavy parasite load.

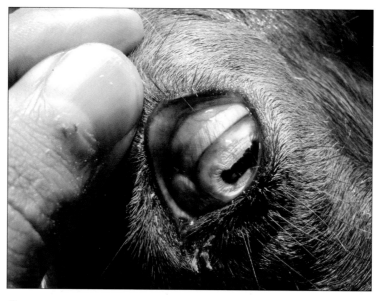

Sheep and goat eyelids should be bright like this.

Internal Parasites, Hosts, and Life Cycles

"Parasite" comes from the Latin word *parasitus* and the Greek word *parasitos,* meaning "a person who eats at the table of another." However, I could not find a record of when the word was actually first used specifically in the biological sense. In 1666, Antonie van Leeuwenhoek invented and used a microscope which magnified up to 200X. He observed and recorded many microorganisms, which he referred to as animalcules (from Latin, animalculum = "tiny animal"). During this time, there was a common belief in *spontaneous generation*, or, of something just "appearing" from nonliving matter. But, Francesco Redi, deemed the "Father of Parasitology," thought differently. His experiments with rancid meat and fly larvae (maggots) proved that the flies were actually laying eggs the eye couldn't see. His use of early microscopes allowed him to view and catalog about 180 microscopic organisms in the 1660s. He too did not use the word "parasite" in describing his discoveries.

Then, in the 1800s the science of parasitology took off. Pierre-Joseph van Beneden, famous for his study on tapeworms in the 1840s, completed a book in 1875 called *Commensals and Parasites in the Animal Kingdom.* Finally, in the early 1900s, parasitology became a recognized course of study at universities.

With grass available, parasites could easily be the reason this goat will not put on weight.

Nowadays, when animals (even humans) become infested or infected with any "bug," "parasite" is often the word of choice. It's used when discussing worms, protozoa, and even bacteria such as E. coli, staphylococcus (staph), and, streptococcus (strep). As well, many species of flies are parasitic, and some whose larvae only are parasitic. So, just what is a parasite? The most recurring definition is: "An organism that lives in or on another organism (its host) and benefits by deriving nutrients at the host's expense."

If your livestock are the hosts, how can you get rid of the parasites sucking their life away?

Well, we're now at the part you will probably have to read at least thirty times: The Classification of Organisms. Classifying organisms helps sort them into groups based on their characteristics, and this aids parasite treatment. Because of shared traits, many parasites within a group can be eliminated with the same substances. To explain the "group" classification, here is a brief lesson, which we all probably forgot after high school, but which is also still under scientific debate and ever-changing. From this book's conception to its publication, the classification (or taxonomy) has changed for several of the parasites listed. Not to open a can of worms, but, here it is . . .

Scientific classification starts with a Super Kingdom, either Prokaryota (life-forms, like bacteria, whose cells have no nucleus or internal membranes), or Eukaryota

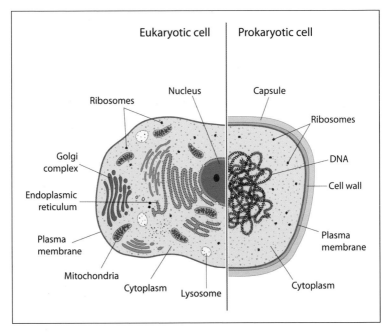

The difference between cells.

(life-forms, like plants and animals, whose cells *have* a nucleus and internal membranes). The nucleus is the control center of the cell, and it's where DNA is stored. Conversely, prokaryotes' DNA resides in the cellular fluid (cytoplasm) and has no membrane around it.

Kingdom comes next in classifying. Currently the United States recognizes seven Kingdoms. The first five are Eukaryotes: 1) Animalia (Metazoa), 2) Plantae, 3) Fungi, 4) Protista/Protozoa, and 5) Chromista. The remaining two Kingdoms fall under the Prokaryote domain, 6) Bacteria, and 7) Archaea. After determining Kingdoms, classification is broken down into Class, Order, Family, Genus, and, Species. Here's an example:

Common Name: Roundworm (human)
Domain: Eukaryote
Kingdom: Animalia
Phylum: Nematoda
Class: Secernentea
Order: Ascaridida
Family: Ascarididae
Genus: Ascaridia
Species: *Ascaris lumbricoides*

"Worm" is another term worth discussing. The scientific name for worm is *nematode*, and there are about 20,000 species, some of which are parasitic. These are called helminths, after the ancient Greek term, "helmins" (meaning, "intestinal worm"). All true worms are in kingdom Animalia, and phylum Nematoda. Knowing this, it's realized that "bots" (an internal parasite frequently found in horses) are *not* worms, but maggots. A "botfly" lays eggs on livestock fur. When an animal uses its teeth to scratch itself, the eggs are accidentally consumed, then hatch and feed inside the animal. The botfly is the parasite, and its young are the larvae. It's not really a worm (nematode) but in the phylum Arthropoda because it's an insect.

One more term needs clarification and understanding: "host-transference," or the ability of a parasite to transfer from one species to another. Remember this fact: *most parasite species are host-specific*. Due to very specialized digestion and assimilation systems in animals, many parasites just cannot withstand *any* body other than its native host. This means the tapeworms in pigs are a different species than the tapeworms in goats. The brown stomach worms in sheep are a different species than the ones in horses,

Internal Parasites, Hosts, and Life Cycles   3

and so on. Here's another example to compare to the previous one (notice only the last classification name, or "species," is different):

Common Name: Roundworm (chicken)
Domain: Eukaryote
Kingdom: Animalia
Phylum: Nematoda
Class: Secernentea
Order: Ascaridida
Family: Ascarididae
Genus: Ascaridia
Species: *Ascaridia galli*

This is important to remember when developing your pasture management system so you can break the cycle of parasites. With the most common livestock, it's sheep and goats that share the most parasite species. Also, many are shared between poultry species: chickens, ducks, geese, and turkeys. In other words, not many parasites will pass from a horse to human, or cow to sheep, or dog to cat, but there are a few. *Ascaris lumbricoides*, the human roundworm, has been found in dogs. Trichinella can pass from pigs to almost any carnivore.

As well, in the kingdom Protozoa there are several species that can transfer between hosts, even to humans. These include the genuses Giardia, Cryptosporidium (or Crypto), and Eimeria. In livestock, both Crypto and Eimeria can cause a disease called "coccidiosis," a broad term which is used because both these protozoa are in the class of Coccidia. These protozoa are shed through feces, and if those feces get into watering tanks, ponds, creeks, or even long-standing puddles, those water sources probably contain Crypto, Eimeria, and/or Giardia. Additionally, many protozoa can live up to four months in ideal water temperatures. Consequentially, the same "watering hole" for cattle and goats (or dogs and chickens!) can be home to many host-transferring protozoa. People should be mindful when handling feces or working around watering systems, and should wash hands immediately afterward to reduce the risk of becoming infected.

Next, it's important to understand the life cycles of a few parasites. Most helminthes mature and mate inside livestock and lay eggs in the animal's digestive tract, which then leave the host in its feces. On the ground the eggs hatch and reside in or nearby the poop to keep from drying out. Larvae can live up to 55

Classic symptoms of "coccidiosis" include pale comb, scrunched-in neck, and puffed-out feathers.

days on the ground before dying, but if the conditions are favorable, some parasites' eggs are able to last through a mild winter. After hatching, the larvae climb up onto foliage. If they are consumed, they live inside their host, develop into an adult, then feed and breed at different locations specific for each particular parasite (mouth, rumen, stomach, intestine, etc.). It will only be 15 to 30 more days before the next batch of eggs are then expelled from these latest consumed parasites. However, several nematodes in the genuses Protostrongylus, Oesophagostomum, and Chabertia live from 30 to 45 days inside the host before their eggs are expelled. (Chart A lists the life cycles of a few helminthes.)

Tapeworms, liver flukes, and meningial worms, on the other hand, need intermediary hosts. With tapeworms, it's fleas or mites who consume tapeworm eggs that will then hatch inside the flea/mite. Livestock inadvertently eat the flea/mite while grazing, or when scratching themselves with their teeth! The flea/mite is digested and the larvae are released into the new host, where they then mature and breed. So the life cycle of the tapeworm can be even longer, up to eight to ten weeks.

CHART A
Some Common Nematodes, Their Life Cycles, and Symptoms of Infestation

Common name	Infectious stage; egg hatches	Usual life expectancy on ground	After ingesting larvae, first eggs seen	Symptoms
barber pole worm	4–6 days	2–4 weeks	3 weeks	swelling under jaw, weakness, anemia
brown stomach worm	4–6 days	several months/ possibly through mild winters	3 weeks	diarrhea, upset stomach, loss of appetite
hookworm	6–7 days	eggs up to 2 years, larvae up to 7 weeks in warm conditions	4–5 weeks	weight loss, diarrhea
lungworm	6–7 days	unavailable	3–4 weeks	sticky nasal discharge, cough, difficult breathing
nodular worm	6–7 days	6+ months	6–7 weeks	dark-green diarrhea
small intestine worm	5–6 days	several months	2–3 weeks	diarrhea, lethargy
stomach hair worm	3–4 days	unavailable	2–3 weeks	anemia, upset stomach, rough coat, diarrhea
threadworm	1–2 days	4 months	8–14 days	weight loss, diarrhea

Flukes, likewise, need an intermediary host, and in this case, it's the snail. Inside a host's bile ducts, adult flukes lay eggs that are passed into feces. Upon hatching, microscopic first-stage flukes migrate, seeking a snail host. They can only survive for a few hours outside the snail, but, once in the snail, they undergo two more stages. Eventually, third-stage fluke larvae emerge from the snail when the temperature and moisture levels are right. These migrate onto forage, where they create cysts and where they remain until eaten by grazing livestock. The cyst is then dissolved by digestion, and the young flukes burrow into the liver, causing considerable damage. About 10–12 weeks after the cysts are ingested and the flukes are mature, the adults move to the bile duct to produce eggs. The whole cycle takes 18 to 20 weeks.

Deer worms, or meningeal worms (*Paralaphostrongylus tenius*), also need an intermediary host, and both snails and slugs (gastropods) will do. First, eggs are produced

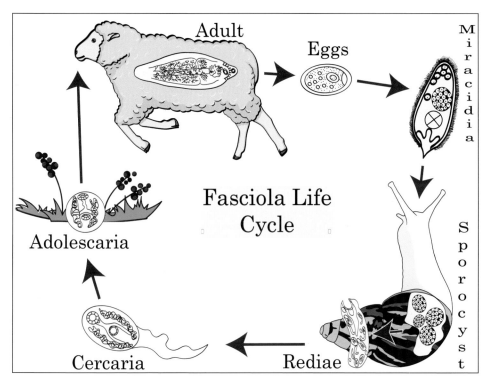

Life cycle of the liver fluke.

by adults residing in the brains of their favorite host, deer. The eggs are carried by blood into the lungs, where they hatch into larvae. Deer cough up and then swallow these first-stage larvae, and then pass them in feces. When a gastropod navigates the manure, those first-stage larvae penetrate into the gastropod's soft flesh. The larvae then go through two more stages inside this intermediary host gastropod.

After three to four months, the third-stage larvae are ready. If the gastropod itself is not consumed, the larvae burrow out of this host, then lie in wait in its slime trail. It is surmised that when either the gastropod or its slime trail gets gobbled up by foraging livestock, they can become infected. The most susceptible secondary hosts are llamas, alpacas, goats, sheep, and calves.

When inadvertently consumed by livestock, the third-stage meningeal larvae burrow out of the digestive tract, then into the host's central nervous system. Research has indicated that the larvae become "confused" when *not* in their normal host (deer, elk, etc.) and most wander aimlessly around the spinal column, never making it to the brain. Irritation of nerve roots at the spine may result in aggravated nerve endings on the skin, and nodules may appear. Many times animals will chew and scratch at these

Threadworms' ideal habitats: 1) Hay lying on the ground around hay rings. At least this farmer has previously moved the ring (as indicated by the "ring outline" in the foreground), 2) In pens where there are thick, gooey layers on the ground!

nodules, producing wounds. Larvae can also create neurological symptoms: wobbliness, rear legs becoming weak or paralyzed, tilting head to one side, and lethargy. Most livestock will present symptoms from 40 to 53 days after ingesting an infected gastropod or slime trail. Also, this parasite can live and produce eggs for up to two years in the brain of a deer, and its first-stage larvae can live up to six months in feces under ideal conditions. However, it cannot survive hot or freezing temperatures for more than a week.

Threadworms (species-specific in cattle, horses, goats, sheep, and pigs) have two methods of infection. The larvae can hatch and be consumed with foraging, or, will burrow into their hosts through their feet! Besides an unthrifty/thin appearance of infected animals, nodules and/or wounds may be visible between the "toes" of cloven-hoofed animals (or in the soft frog of horses). The most common causes of extreme threadworm counts are 1) excessive stocking rates of livestock, 2) unclean bedding or sleeping areas, and 3) standing for lengthy periods in infested areas, like around feed bunks, hay rings, or dairy barns at milking times.

Another parasite requires the acid and pepsin of its host's digestive system to break down its "shell" (cyst). Then, when the newly hatched larva arrives in the small intestine, it can burrow in, mature, mate, and release *live* larvae into its host's system, all within a week! What is this speedster? *Trichinella spiralis* (which causes trichinosis)

Muscle tissue with encapsulated *Trichinella spiralis*.

is the smallest host-transferring, parasitic nematode. It lives most its life encapsulated in a cyst in the muscle tissue of carnivores and omnivores. It is passed into new hosts when infected meat is eaten. This is why it's so important to properly and sufficiently cook pork, to kill the parasite's larval cysts!

After arriving in the intestine, it is only about a week before a female is mature enough to mate and create up to 1,500 *live* larvae. Upon being born, the larvae migrate into the bloodstream, and then into muscle tissue, even of the heart. A female's life is about six weeks, a male's life is even shorter, and this is important because it's the only time this nematode and some of its offspring can be killed or ejected with "dewormers."

And, yes, pigs *will* eat meat, mostly as carrion. Pastured pork, with plenty of access to fields and forests, will eat dead animals (or kill live, injured ones!), which have been infected with trichinella, such as rats, foxes, raccoons, possums, skunks, and even predatory birds.

Common Internal Parasites of the Most Common Farm Livestock

Cat: Coccidia, hookworm, roundworm, whipworm, tapeworm

Cattle: Coccidia, bankrupt worm, barber pole worm, brown stomach worm, small intestine worm, nodular worm, hookworm, hair worm, liver fluke, lung worm, tapeworm, threadworm

Dog: Coccidia, giardia, hookworm, roundworm, whipworm, and tapeworm

Fowl: Coccidia, hairworm, roundworm, threadworm, gapeworm, gizzard worm

Goat/Sheep: Coccidia, barber pole worm, bankrupt worm, brown stomach worm, lungworm, nodular worm, small intestine worm, threadworm, tapeworm, meningeal worm, liver fluke

Horse: Strongyles (red worm/blood worm), small strongyles, large roundworm, threadworm, pinworm, bots

Pig: Eimeria, large roundworm, whipworm, nodular worm, threadworm, kidney worm, lungworm

Rabbit: Eimeria, giardia, stomach worm, pinworm, tapeworm

External Parasites

Most external parasites are pests we can see, and sometimes we see way too many of them! Even if internal parasites seem to cause more actual deaths in livestock, external parasites affect animals' growth and production, and can carry diseases that could result in death. At a minimum, external parasites plague animals relentlessly, causing weight loss and unthriftiness. So, reducing lice, mites, flies, grubs, keds, fleas, and ticks can result in greater life expectancy and greater yields from livestock.

External parasites stress animals, cause dairy livestock to reduce milk production, and make meat-animals gain less weight because their time is spent trying to keep these pests off instead of eating. These reductions cause economic losses. Horn flies (the small gray fly found on the backs of cattle all summer) cause an estimated $1.36 billion in annual losses to cattle owners in the United States. Horseflies create another $672 million in losses, and face flies $192 million. Even ticks have caused $162 million in losses, with the common dog tick leading the pack. One prevalent tick- and fly-transmitted bacteria is *Anaplasma marginale*, which causes anaplasmosis

Horn flies plague cattle, causing weight loss and unthriftiness.

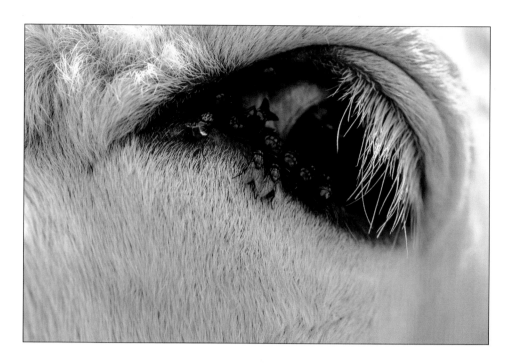

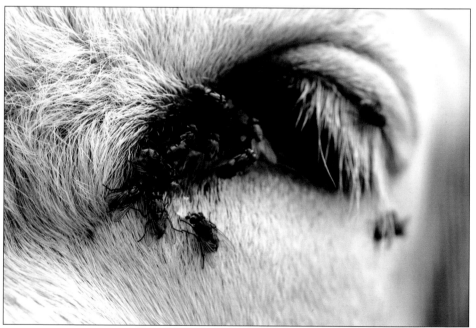

Flies swarm to livestock's eyes. Wounds can develop, increasing fly activity and infection.

in all mammals, including humans. Anaplasmosis results in severe anemia and weight loss. Both biting and/or blood-feeding pests are able to pass on this bacteria from animal to animal, spreading it quickly through a herd. Even the nonbiting face fly can cause and rapidly pass along pinkeye (infectious keratoconjunctivitis) from one animal to the next, after ingesting fluids from livestock's eyes and transmitting the bacteria from their saliva as they "drink." There is even a fly whose young can live inside its host's digestive tract.

Most seasoned horse owners know of the botfly and have seen its minuscule yellow eggs on their horses' front legs. When a horse uses its mouth to chase away biting flies, its spit causes the eggs to hatch. The next time the animal chases a fly with its mouth, the larvae get onto the horse's lips, and are subsequently swallowed in the next mouthful of hay or pasture grass. They then attach themselves to the lining of the horse's digestive tract, getting their food and oxygen from their host. After about 8 to 11 months, the larvae are eliminated in the horse's manure, where they pupate and 4 to 8 weeks later hatch into adult botflies.

Sheep and goats are often hosts to the nose botfly. Live, active larvae are deposited by adult botflies in the nostrils of these ruminants and can cause a condition called "blind staggers," because they will stagger and/or toss their heads as if blind. A runny

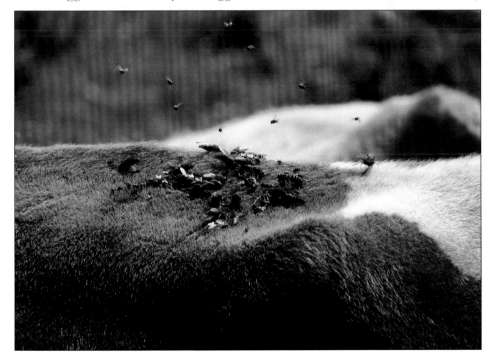

Flies can create large wounds, ruining hides.

nose, shaking their head, blowing sharply from their nostrils, and subsequently not eating as much can reduce weight gain up to 5 percent, and milk production up to 25 percent! For wool producers, the fly that can break the bank is the wingless sheep ked. Though they look more like a tick and are often called "sheep ticks," they are flies. Reddish to gray in color, these ¼-inch-long bloodsucking pests quickly multiply and spread though herds. Keds affect wool production by decreasing their hosts' blood supply and thus their hair length. Additionally, keds' excrement stains the wool, and pupae found in the wool reduce both the condition and the price of the wool.

Biting flies also create wounds and sores on the skin of cattle, goats, and other hide-producing animals, reducing the value of these hides. Deer- and horseflies are particularly vicious, creating large, irritating wounds. Livestock scratch on whatever is available, often removing hair and causing thick calluses on the skin. As well, horseflies can incite livestock to run, usually in the hottest time of year. In a fly's prominent eating/breeding season this can result in stressing livestock and a reduction in weight gain and/or milk production.

Constant aggravation can lead to running and rolling.

Other external parasites can also damage the skin, hide, hair, or wool of mammals, and the skin and feathers of birds. The bloodsucking of lice, mites, grubs, fleas, and ticks causes itching, leading to scratching, biting or pecking, hair or feather loss, and thickening of the skin or hide. This activity is usually heaviest in winter or early spring when the host has more "insulation" (hair or feathers) for the pest to hide under and keep moist. Besides causing heavy scratching from the bite, the pests (and some parasites' eggs and larvae), also create lower-quality fur and wool. Infestation of certain species of mites in mammals is called mange, and a very aggressive type of mange is scabies, caused by sarcoptic, chorioptic, and/or psoroptic mites. Scabies is serious, and must be reported to disease control authorities. Many states require auctioned

Lice can cause itching and heavy hair loss.

Mites can cause mange/scabies.

animals to have an identifiable tag or a tattoo to be able to trace where severe cases originated. Lastly, heavy lice and/or mite infestations can greatly affect egg production in layers.

As if that's not enough aggravation, fleas and mites can transmit tapeworms to livestock. If you refer to the earlier chapter on internal parasites, you'll remember that fleas and mites will consume tapeworm eggs. When the flea or mite feeds on livestock and that animal then bites at the itchy area, it inadvertently eats the parasite, whose eggs will hatch inside your animal.

You're probably itching to know by now just how to eliminate these pests. External parasite control can be accomplished by dipping, spraying, and/or dusting livestock, and also through environmental control. Most natural treatment mixes will need a carrier powder (for dusting) or a carrier oil/mixture (for dipping or spraying). Sulfur, potash, and diatomaceous earth (DE) are powders frequently used in dusting mixes, all which contain antiparasitic properties. Sulfur kills and/or repels many external parasites, whereas potash (literally sifted from fireplaces) and DE. both act as desiccants, drying out both the larvae's and adults' exposed soft bodies. Also, dusting creates a dry "environment" that prevents parasites' eggs from hatching. Something that truly puts a smile on my face is knowing that DE actually "scratches" and perforates the

There are many kinds of dipping tanks.

soft, waxy thorax of lice, mites, grubs, and fleas, creating microscopic lacerations! Sadly, though, ticks are not affected by DE or potash; their hard, slick bodies repel powders. A carrier oil is necessary to soak their outer shell and deliver the natural pesticide internally, with the oil also "suffocating" the parasite because its skin cannot breathe. Fortunately, many dips/sprays will also penetrate soft-bodied pests.

You may wonder how you are going to "dip" a cow, or spray or dust 100 head of goats? Dipping livestock does require a walk-in dipping tank, and spraying or dusting will require a squeeze-chute/ stanchion. Additionally, lice and mite treatment will need a second application in one to two weeks to kill any newly hatched parasites. As well, fly, tick, and flea control will need routine, once-a-week application (twice a week is ideal) during the heavy season (most of the summer and early fall). So, having a dipping tank or head-chute/stanchion will be an investment worth undertaking.

I have also done a "modified" dipping using only a head-gate. First, I create my solution in a five-gallon bucket(s). I don forearm-length heavy-duty rubber

Classic squeeze chute for cattle.

gloves, a moisture-resistant jacket, and a pair of goggles. With a large, thick sponge, I then apply the "dip" solution onto the animal. I use almost the entire five gallons on one cow or cow-sized animal, but, luckily, the majority of the dip is warm water. This process is not very successful with heavy-coated sheep. If you are a hair/wool-producer, a dipping tank truly would be a great investment.

Ingredients for External Parasite Removal

Here is a list (and a brief description) of products which meet the US National Organic Standards for external parasite removal:

1. **Soap:** Any pure, environmentally safe soap will work by removing the waxy cuticle that protects soft-bodied parasites from drying out. A one- to two-week repeat application will be necessary.

2. **Oils:** Organic (or natural) plant or nut/seed oils work by clogging up the pores that deliver oxygen to the parasites' bodies. I've used sunflower, olive, and pumpkin seed oils. I've even used my expensive cooking oils when they've gone rancid. You can simply apply a light coat of oil to the infested area, or run oil along the neck and spine to cover the most commonly infested areas. Do not overdo the oil, as a thin, even application will work. Just as with soap, repeat treatments are necessary. Also, do not use partially or totally hydrogenated oils, as animals may inadvertently ingest them. And *never* use kerosene or petroleum-based products like diesel. Mineral oil is the only USDA-approved petroleum carrier oil.

3. **Liquid enzyme spray** works by breaking down the exoskeleton of parasites, and hence, drying out the little buggers. Be sure these are natural enzymes derived from edible, nontoxic plants, are not genetically modified, and are NOT derived from infective bacteria or fungi which cause diseases. (I have not personally used enzymes.)

4. **Diatomaceous earth** (food-grade, *not* the kind sold for pool filtration). As mentioned above, it slices soft-bodied parasites and is also a desiccant. *Use caution when applying the powder; wear a mask and avoid kicking up clouds of the dust which can enter the eyes and breathing passages of animals and/or humans.*

5. **Garlic powder** has an active ingredient called allicin. It can kill or repel parasites and also has excellent antimicrobial properties. (Feeding garlic in conjunction with external treatment has shown promise, especially with mange mites.)

Here are a few USDA Organic Restricted Use Products, which, if "certified organic," may be effective, but you may need approval from your USDA organic certifier before using them.

6. **Neem oil** comes from the neem tree of India, and this natural insecticide is effective against all external parasites. At least a 1 percent solution must be used, but studies have shown that up to a 10 percent solution is safe for

animals. I have personally used a neem concoction (described later) for lice, mites, and fleas, but I never sought organic certification.

7. **Pyrethrum (chrysanthemum flower)** is a true botanical insecticide that kills parasites on contact, and is best used only for heavy infestations. Make sure the formulation you purchase does not contain piperinyl butoxide, which is prohibited for use in certified organic operations. Pyrethrum works great as a fly repellent as well. I've used it extensively on my horses and family milk cow.

There are also a few carrier oils that have the extra benefit of being antiparasitic: pumpkin seed, grapefruit seed, and grapeseed. These are usually more expensive than oils like sunflower or even olive oils, but I've used pumpkin seed oil after it has gone slightly rancid and could not be used for cooking. Additionally, using small amounts of certain essential oils in a mixture can add even more insecticide-boosting power. Try adding just a few drops of juniper, pine, eucalyptus, lemongrass, clove, thyme, oregano, or tea tree oils for every 2 tablespoons of carrier oil. Most of these also possess antifungal and antibacterial properties, helping to decrease infections from open bites or torn flesh from rubbing or scratching. Cats are extremely sensitive to many essential oils, lacking the proper enzymes to break down and eliminate them. As well, chickens show hypersensitivity to essential oils. Because most cats groom themselves regularly, ingesting essential oils can be fatal in high doses. Using only 3 drops of essential oil per 1/2 cup of a carrier oil (or per mixture described below) is sufficient. Pets or livestock should never receive treatment with undiluted essential oils, externally or internally. *As a general rule, if an essential oil is harsh or harmful on human skin, it will be for animals as well. Read labels and do research before using any ingredient.*

External Parasite Mixes/Concoctions

Fly-Repellent Spray—For every gallon of warm water, add ⅓ cup of a carrier oil, ¼ cup of soap, 1 teaspoon of tea tree oil, and/or 40–50 drops of oregano and/or pine essential oils. I've also just followed the ratios/instructions for pyrethrum (ordered online) and have added essential oils, even citronella! But, remember, if you're a certified organic operation, get approval from your certifying agent to use pyrethrum on livestock are in your organic operation.

Research shows use of essential oils can be from .5% (3 drops per ounce of carrier oil) to 10% (60 drops per ounce of carrier oil) of an external parasite mixture, depending upon the essential oil. Before adding essential oils to your concoctions, consult an experienced aromatherapist and/or do additional research on any essential oil(s) you wish to add. And, use extreme caution when developing a mixture for cats as they are highly sensitive to these oils.

Lice, Mite, Ked, Flea, and Grub Dusting Powder—Use equal parts of each: sulfur powder, DE, and potash. For every full cup of dusting powder mix, you can add 1 tablespoon dried herbs like oregano, thyme, rosemary, etc. I have used this on goats, chickens, and our dairy cow. Wearing gloves and a face mask, dust onto animals' fur or feathers. With a comb or brush, work the powder next to the skin.

Lice, Mite, Ked, Flea, Tick, and Grub Spray or Dip—For every gallon of warm water, add ⅓ cup of a carrier oil (see page 19), ¼ cup of soap (see page 18), 1 tablespoon to 1½ cups (same organic certifying rules apply) neem oil*, and/or 40–50 drops of oregano and/or pine essential oils. I have used this on chickens, goats, dogs (as a dip), and cattle. Stir or shake well before use. When using a walk-in dip bath, be aware that the oils settle on the top of the water and they alone may need to be added in lesser amounts periodically to the remaining water.

I use 1 tablespoon of neem oil for "preventive" or very light infestations, and 1½ cups of neem oil for heavy infestations. For moderate-sized populations of parasites, ¾ cup neem would suffice.

Traps—Because of my concern about environmental control, I have not experienced such major infestations of flies to warrant building/using traps. However, there are many, many designs online. ATTRA (Appropriate Technology Transfer for Rural Areas) has a good overview of organic control of parasites in cattle and it discusses a few traps and other environmental controls. (See page 80, Resources). I've also researched horse- and deerfly traps and watched videos showing the flies diving right into a death trap. Marvelous! This particular trap was quite large, maybe four or five feet tall by three feet long, and two feet wide. Like a feed trough, it has a tray halfway between top and bottom. There is a V-shaped "ramp" leading from the top of the trap to the tray, with a three- or four-inch gap at the bottom of the ramp which feeds into the tray. There is a liner in the tray which is filled with soapy water. Flies "think" this tall trap is a large animal and, when they go to land on it, they slide into the tray of soap water. The soap coats their wings so they cannot fly, and they soon drown! This has to be set up in an area *not* directly accessed by livestock, but should be placed near where livestock congregate regularly. I would recommend building a four-sided cattle-panel "pen" for it to keep livestock from rubbing against it and knocking it over.

More on environmental control of external parasites will be covered further in the "Pasture Management" chapter ahead.

Other Disease-Causing Organisms

The focus of this book is *not* on bacterial or fungal infections. However, in the course of doing research for this book, contemporary articles about parasites were found that would include information on fungal and bacterial infections, since animals with external parasites (and a few internal parasites like menengial worms or nodular worms) can create external lesions that then become infected. Common fungi like aspergillus (mistaken sometimes for brucellosis, which is caused by a bacterium, brucella), the fungi "ringworm," and other fungal infections can cause problematic diseases in livestock. Consequently, because these infections are usually in conjunction with a parasite infestation, and because they can have such a dramatic influence on livestock, I have included a chart (chart B) of the most common plants with antibacterial and antifungal properties, and the substances responsible for these properties.

At the beginning of chapter 1, the words "prokaryote" and "eukaryote" were mentioned. These terms define organisms with a certain cell structure. Eukaryote cells have a true nucleus (and membranes surrounding their organelles). Most multi-celled organisms are eukaryotes, including humans and livestock. If even a single-cell organism has a true nucleus,

Fungal disease on a hen's comb, eye, and cheek.

it too is a eukaryote. Fungi are eukaryotes. Prokaryotes, on the other hand, do not possess a true nucleus or membranes around the organelles. Bacteria are prokaryotes. Salmonella, E. coli, staphylococcus, and streptococcus are the most common prokaryotes. Usually, they are much smaller than eukaryotic cells; most bacteria are not visible until 400X magnification. Since affordable microscopes only possess 10X, 40X, and 100X magnifications, it's impossible to identify these organisms at home, so it's important to have a veterinarian to diagnose the problem correctly.

But, as a first step in controlling the bacterial and/or fungal disease, it is recommended that affected animals be segregated and their pens or stalls cleaned and disinfected. Unaffected livestock that have been in contact with the diseased animals should be watched closely for the appearance of lesions and be treated promptly. Remember that flies are still capable of carrying infections from the affected animals to the other, healthy livestock, so the use of fly traps may be beneficial around the segregated animals. After you and/or your veterinarian determine if the cause of a problem is bacterial or fungal, you can use that information to research specific treatment protocols and dosage amounts of natural treatments. Though research indicates both neem and tea tree oils fight fungal infections, you will need to determine specific dosages and treatments for your animals' fungal and/or bacterial infections.

CHART B
Common plants with antibacterial and antifungal properties

Plant	Antibacterial Substance
anise	estragole
clove	eugenol, tannins
cinnamon	eugenol, tannins
garlic	sulfur, geranium
nutmeg/mace	eugenol
onion	sulfur
oregano	thymol, carvacrol

Pasture Management

First and foremost, to reduce the parasite load on the ground, it's essential to develop a working pasture rotation system. Whether you a) have single-species livestock, b) have a multispecies grazing program, or c) use leader-follower succession grazing, it is very important to have a solid pasture rotation system. This is done by manipulating both 1) the length of time for grazing, and 2) grazing habits.

After parasite eggs are expelled from the host in poop, most will hatch in 1 to 14 days, with the average being about 5 days (Chart A, page 6). Then, many species of larvae can live up to 40 days if conditions are favorable—moist with moderate to warm ground temperatures. Most parasite larvae will be dead by 55 days, *if* they are not consumed by a host before then.

By understanding parasite life cycles, you can adjust the number of grazing days to ensure livestock are not reinfected by the same parasite species. For instance, in favorable conditions, in as few as *four* days barber pole worm eggs hatch and can reinfect your goats or sheep. Because larvae only survive on the ground for about four weeks,

Easy-rolls like this allow varied lengths of electric cross-fencing to be moved and adjusted quickly.

if that parasite is causing considerable damage, it would be prudent to move goats or sheep every four days (or more frequently) during high-load times. Then, let those sections rest from goat/sheep grazing for at least 60 days (to ensure *all* parasite larvae have died). This can be accomplished by further subdividing pastures with electric or temporary fencing.

With a same-day hatch-rate, threadworms are particularly difficult to kill off. An organic dairy farmer I interviewed who was having an issue with threadworm spent six months moving cattle *twice a day* into new subdivided sections. He then did not allow cattle back on the fields for almost a year. In just six months (and no chemical antiparasitic treatments!) his diligence paid off and the threadworm count dropped to almost nonexistent!

Manipulating grazing habits is also important when designing your rotation system. Livestock could be grazed according to the pasture's orientation and/or the plants that could benefit the livestock during a certain season. Spring pasture could be sloping and south-facing to allow moisture to run off and for the sun to do its magic more efficiently on the larvae. In wet grass, larvae can be found moving up to a foot away from feces. However, they venture only a few inches away from the ground when the grass is dry. Also, larvae make their way higher up onto the plants when the

Move livestock frequently into pasture sections divided with one or two strands of electric fence.

sky is grey and overcast, like when it rains or in the early morning and late afternoon. They avoid bright light.

Summer pasture could be lush lowland grazing areas because at this time of year the hot sun forces larvae closer to the ground and moist manure. Grazing animals will consume fewer larvae. Then, fall pastures might be areas with more nuts, fruits, and leaves available. Goats, sheep, hogs, cattle, and even horses appreciate acorns and other fallen treats. Just remember to monitor the moisture levels and temperatures of fall pastures. Many parasites will be able to survive under leaf cover. Finally, in USDA Hardiness Zones

A south-facing, sloping pasture works great in spring.

Summer pastures are lush lowlands, with plenty of shade.

of at least a 6 or higher, winter pastures could contain taller, "stocked" forage. Parasite eggs and larvae do not last long in cold winters, even if the forage is tall.

After pasture rotation, the next most beneficial step in parasite control is proper environmental maintenance, creating healthy pastures and clean watering systems. Unless you have the perfect stocking rate of livestock, you will probably have

Pigs help to break up the manure of other livestock, exposing parasite larvae.

to mow pastures to keep them producing better forage and to reduce parasites. But if you're striving for more sustainable practices, "wasting" gas (and not utilizing the free grass) seems silly. So, during wet, heavy-growing periods it is wise to utilize multi-species grazing or leader-follower grazing.

Leader-follower rotation: After goats (background) have had their fill, pigs will move onto the pasture in 2–3 weeks.

Since most parasites are host-specific, cattle and goats or horses and sheep could join or follow one another. (Remember: sheep and goats share many parasite species and should not be grazed in succession of each other.) Also, during heavy growing periods, be careful not to rush livestock *back* onto pastures before 60 days. This can increase parasite levels. If this is not possible, you can lower the rate of infection by *only* grazing livestock on such pastures during dry, bright, daylight hours—after dew has evaporated, or the grass has dried after rain.

Also, avoid grazing or cutting pasture grass below four inches, especially during wet or dark periods, like spring and early fall. If grass is short there is a higher chance that grazing animals will consume more larvae since the parasites' larvae won't have far to climb to be ingested. However, leaving grass too long (over 10 inches) will not allow for enough sun to dry up eggs and larvae. Many resources indicate that pasture grasses should be kept around 6 to 8 inches for optimal forage and parasite control.

This pasture has been grazed a bit low, forcing animals to eat too close to the ground. But multispecies grazing helps to slow reinfestation.

Healthy pastures also need beneficial soil organisms such as predatory nematodes, ground and dung beetles, and certain fungi. These organisms attack parasite eggs and larvae, and/or keep manure dispersed. Scattered poop dries more quickly and leaves very little opportunity for parasite larvae to keep moist. Even fly populations can be reduced up to 90 percent by these soil improvers.

Predatory/beneficial nematodes (remember, nematode means "worm") destroy over 200 species of eggs and larvae, many of which are plant parasites, like the corn cutworm. Several species of these nematodes are used in organic agriculture. There are also predatory/beneficial nematodes that prey on insect and parasite eggs or larvae. *Steinernema carpocapsae* is most effective against larvae of houseflies and stable flies, and it also hunts down flea larvae and pupae (secondary hosts of tapeworms!). Other fantastic predatory nematodes are in the genus *Butlerius*. Several species have shown to greatly reduce parasite larval populations. In one study *Haemonchus contortus* (barber pole worm) was reduced by 61.9 percent!

Other pasture health-boosters include a variety of beetles. Predators like ground beetles (family Carabidae) and rove beetles (family Staphylinidae) feed on larvae of flies and parasites. Dung beetles play a huge role in parasite reduction too. There are

Nice color and shape for goat or sheep manure.

In healthy pastures/paddocks (with very limited chemical usage) manure is broken down and dispersed in days.

This cow pie is a bit firm, but of good color.

After a few days, dung beetles have done their work!

three types: 1) rollers/tumblers (such as *Canthon pilularius*), 2) tunnelers (*Onthophagus gazella* or *O. nuchicornis*), and 3) dwellers (the Aphodiidae family). Tumblers and tunnelers are the most disruptive to parasites because they bury manure in the ground for their own eggs, sometimes breaking up an entire dung pat within 24 hours. This affects the parasites' eggs, which become trapped too far underground. Hatching larvae cannot make it to the surface and will die. Plus, the beetles' holes and the buried dung increase soil and plant health. As well, livestock manure is more broken up and dispersed, and dries more quickly, killing any remaining parasite eggs or larvae.

However, if a lot of manure becomes "spread out" it could also make ingesting larvae more probable, since livestock will find it difficult to graze away from the scattered feces. Larvae can wriggle up to a foot away from poop in wet, overcast conditions. So, if your pasture or paddock seems to have manure that gets broken down within a few days, think seriously about moving animals more quickly from those sections until the spread manure dries out. Likewise, you may want to prohibit your free-range chickens or hogs from multispecies grazing with cows or other large livestock (and spreading their manure!) until *after* you have moved that livestock to a new area.

Lastly, researchers are studying parasitic fungi for control of livestock parasites. Two types, *Duddingtonia flagrans* and *Verticillium chlamydosporium*, have spores with thick outer walls, which prevents them from being broken down in a digestive system. After being consumed by livestock, then excreted in the poop, the spores hatch and "attack" parasite eggs and/or larvae.

If your pastures lack beneficial mini-predators, there are companies which sell just what you need. At the time of the writing of this book, several species of beetles and predatory nematodes were available. However, parasitic fungi were only available to purchase for research. To help beneficial mini-organisms to flourish in your pastures, it's necessary to develop healthy "living soil" by following these steps:

First, reduce close-cut mowing or haying and/or repeated strip-grazing, since constant exposure of the sun on the soil will not allow for moisture-needy organisms to flourish (remember the 6- to 8-inch growth rule). Though moisture is beneficial to these mini-predators and their young, it is also exactly what many troublesome parasites need. Keeping grasses and shorter forages around 6 to 8 inches will allow animals to eat farther from the ground, away from parasite larvae and eggs (and intermediary hosts like snails/slugs, fleas, and mites!) who stay closer to moist earth and/or manure.

Second, do *not* use chemicals of any kind on your pastures—no chemical fertilizers, herbicides (plant killers), or pesticides ("bug" killers), including chemical

Keep cattle from standing (and pooping) in ponds.

Pigs will want to get into water tanks with low sides. Use smaller tubs and clean and fill at least twice a day, or, use a water tank with drinking nipple for hogs.

dewormers and even chemical livestock sprays, dips, and powders. Research indicates these chemicals kill and/or disrupt the life cycles of beneficial organisms. In one study, dung insects on conventional farms were diminished over 50 percent compared to organic farms. If you need to resort to chemical compounds, limit your livestock's range to a "sacrifice" area so their chemical-laden feces are also limited to that one area. The most harmful chemical antiparasitics to soil organisms are those ending in "ectin" (like Ivermectin and Cydectin), and those ending in "thrin" (like Deltamethrin and Permethrin).

Environmental maintenance also includes promoting clean water systems by reducing long-standing, high-moisture areas. Protozoa infestations (Giardia, Eimeria, and Crypto) can be minimized, as well as snail and slug populations. This is accomplished by 1) reducing standing-water puddles with fill-in, 2) keeping livestock from drinking and standing in "natural" sources, like ponds and creeks, or limiting their grazing time in pastures with natural water sources, and 3) compelling livestock to drink from tanks that can be kept clean regularly. These measures will reduce protozoa populations, and without constant moisture, gastropods cannot survive.

Movable tanks with floats like this are easy to move from section to section when rotating frequently.

Clean water is paramount for great health!

To also keep livestock from ingesting gastropods/slime trails (which can result in meningeal worm), it would be sensible to deter deer from pastures where you graze the most susceptible livestock: goats, sheep, llamas, alpacas, and calves. Top-strand electric fencing works somewhat well, but, if the deer doesn't touch it on its first jump, it will continue to jump and graze. Guardian dogs work relatively well too, but, if they are guarding the livestock, deer can infect any empty/resting pastures. So, before the livestock enter a "new" pasture, you might consider a preemptive strike using ducks, chickens, and other fowl to get rid of gastropods. Or, if you are aware of areas on your farm with high deer populations, it might be wise to just keep susceptible livestock out of them until hotter/drier weather.

The final steps in environmental maintenance are to create 1) a sacrifice area and 2) a beneficial plants paddock. As mentioned earlier, sacrifice areas (dry lots, paddocks, pens) are used when chemicals need to be administered to livestock, or when pastures desperately need rest. If livestock remain in such an area for more than 30 days, it is imperative to administer antiparasitics, or provide antiparasitic free-choice mineral licks or fresh-cut antiparasitic herbs (see Eliminating Parasites with Antiparasitics, page 34). Livestock in confinement often pick up *any* morsels from the ground, and could consume parasite eggs. So, keep manure picked up. Less poop and drier lots equal fewer parasites.

Beneficial plants paddocks could also reduce parasite loads up to 50 percent. Plants high in tannins—such as chicory, dock, plantain, raspberries, bird's-foot trefoil, *Sericea lespedeza*, and oak—are nematodes' enemies. Running livestock through these "stocked" pastures during high-load times (spring and fall) will improve animals' health. Besides tannin-based plants, there are many other native plants and herbs that can also be grown in these pastures (located in the "List of Antiparasitics," pages 39–72).

Examples of Pasture Rotation

Following are a few examples of rotation systems.

Right after kid-birthing takes place in April, one farmer runs her goats and their young through a beneficial-plants paddock (under one acre). Five to seven days later they move to their first pasture, and 35 days later into a second pasture (each about two acres) where they graze for another 35 days. In the summer she keeps them on a five-acre pasture (south sloping, ½ treed) for three months. Barber pole worm, especially, can be greatly reduced because of its low life expectancy on the ground and intolerance of dry, hot conditions. Then, in the fall, she moves them again through the beneficial plants paddock, and five days later moves them to a one-acre, sloping pasture for the moist, fall season. She chooses to keep them through the winter (December through March) in a large sacrifice paddock, feeding hay and other fodder grown through the year (beets, turnips, dried greens, and collected nuts). Not only is this done to rest the pastures, but coyotes and mountain lions are more active during winter in her area. She then starts the system over in the spring after kidding.

Another person runs their family dairy cow on small, ¼-acre plots for 1 to 2 weeks each, getting her through six sections within 70 days. In a few of those sections they also move poultry daily over the ground in large tractors to break up manure, especially in July and August. They offer the cow handpicked beneficial plants all throughout the year, and some sections have more varied species of plants for the cow to self-select as well. When she dries off she is moved through two larger pastures after 4 to 5 weeks each, following meat goats, then starts back on the ¼-acre plots.

Yet another family utilizes large pastures with electric wire to section off even smaller areas where 80-some dairy cows remain for only a day or two at a time. They never graze the same spot until the next year. Similarly, I have done this with goats and chickens using electro-netting, never grazing them on the same spot for almost a year.

Within the Pasture Management guidelines (see page 23), it's up to you how to best graze your livestock to allow for high nutrition intake, to reduce the parasite load, and to work with the time and space you have to offer.

A Final Word on Environmental Control of Parasites

Be sure to keep paddocks, stalls, feeding areas, coops, and high-traffic areas clean of accumulating muck and damp debris such as loose hay, leaves, etc. These act as protection for parasites. Something as harmless as an animal nipping at its ankle could result in ingesting parasite larvae. Clean out from under hay racks regularly so animals aren't tempted to nibble through the moist remains and pick up larvae.

Another hint is to use barn lime, sulfur powder, wood ash, or diatomaceous earth (or any combination of the four) after each mucking to dry out larvae and eggs that remain even after cleaning. Then, between cleanings, sprinkle the powder in the animals' bedding, or in nesting boxes, to help keep parasite levels lowered. Even using orange oil or neem in a garden pump-sprayer will help kill parasites and larvae.

A good rule of thumb is to follow a three-day parasite treatment with a thorough barn cleaning and a dusting of the stalls and animals' "resting areas" with any of the four drying agents. Then, offer free-choice herbs (beneficial paddock, fresh, or in a mineral-powder mix) for another three days.

Eliminating Parasites
with Antiparasitics

After the pasture rotation and maintenance systems are running smoothly, the next step is to rid the body of any parasites that have made it in anyway. Parasites of all kinds live in our world and it's almost inevitable that all animals will get them at some point. All wild animals, and very often domestic animals, live with some kind of parasite(s). It may not be visible in livestock who are well cared for, because healthier animals are more resistant to the *effects* of parasites. Resistance levels vary from animal to animal, but nutrition and general care can raise, or lower, their resistance.

Rotation is your first defense, but, if your livestock return to the same barn, hut, or hayrack to eat feed or hay (like most fowl, pigs, and dairy animals) it's imperative to also keep infestations low in those areas. Simultaneously cleaning these homes *and*

Sadly, this kid will most likely ingest parasite larva from nursing off his dam's dirty teats. The diarrhea on the doe's legs and tail are indicative of parasites.

deworming before rotating livestock will keep parasite eggs from being dropped with the manure in a new area. This is particularly important during high-load times like spring and early fall when moisture increases parasite life expectancy.

To get rid of these little buggers, we have to administer an antiparasitic (also known as an anthelmintic, antihelminthic, or "dewormer"). Many natural ingredients possess antiparasitic properties. Some of these are vermicides, which kill the parasite, while others are vermifuges, which only numb or knock out the parasite so it can be easily flushed from the body. Chart C lists some of the natural compounds that work as antiparasitics. This is only a partial list and, with a little research, I'm sure you will find more to add.

Natural dewormers work differently, so some require daily doses to get rid of parasites, and some require regular, intermittent doses. Please read the "List of Antiparasitics" (pages 39–72) carefully, and choose the one(s) that are prevalent or easy to acquire, and ones that work with your rotation schedule.

Also, there is a lot of information about natural and herbal remedies on the Internet, but much of it is anecdotal. Very few scientific studies have been conducted in the United States regarding the use of herbal treatments, and much of our nation's livestock ingests chemical antiparasitics regularly. However, through historical documentation, and through researching international studies, I was able to discover the secrets to ridding parasites from our livestock *naturally*.

CHART C
Natural Antiparasitic Compounds

Family (if mentioned)	Ingredient	Found in
Artemisia	santonin, coumarin, thujone	tarragon, wormwoods, angelica
Apiaceae/ Umbelliferae	limonene, fenchone, lactones, sulphur	fennel, carrot seed, parsnip, anise, coriander, parsley, dill, cumin
Brassica		mustards, radish, turnip, nasturtium, rutabaga, kale, collards
Cucurbitaceae*	cucurbitin	squash seeds
Labiatae	thymol	thyme
Tanacetum/ Asteraceae	terpenoids, thujone, sesquiterpene	tansy

Fabaceae	alkaloids	lupines
Ranunculacece and Beberidaceae	berberine	goldenseal, barberry, Oregon grape, yellow root
numerous families	tannins	*sericea lespedeza*, dock, birdsfoot trefoil, black walnut, oak/acorn
seven families	terpene	conifers

*The Cucurbitaceae family (squash) possesses an amino acid called cucurbitin. It's the active compound that eliminates worms. This amino acid is found only in the raw (not roasted) *seeds* of the squash. However, the concentration of this substance varies in each species, and some families of Cucurbitae tend to have a greater concentration of cucurbitin.

> *C. maxima* contains 5.29% to 19.37% cucurbitin
>
> *C. moschata* contains 3.98% to 8.44%
>
> *C. pepo* contains 1.66% to 6.63%

Currently, a source could not be located which included the *C. mixta* family, nor one which lists the individual species' amounts of cucurbitin. Suffice to say, if you're shooting for "maximum" effect, your best bet is to go with the *C. maxima* family. You can find out what family a squash or pumpkin is in by searching garden seed sites online, or in gardening and seed-saving books. Your local greenhouse or garden shop may also have that information. Here is a short, incomplete list of some of the families' species.

C. maxima	*C. moschata*	*C. pepo*
Amish pie pumpkin	Butternut	Acorn
Atlas squash	Cheese squash	Cocozelle
Banana squash	Cushaw (ONLY gold & orange)	Crookneck
Buttercup	Maryland pie pumpkin	Gourds
Hubbard	Virginia Mammoth	Scallop
Turban	Mallow	Zucchini

Dirty tails and hind legs are stark indicators of high parasite loads.

If this sheep would have had normal, pelleted manure, these dangling diarrhea balls wouldn't have accumulated to this degree. Under the wool is the perfect place for other, external parasites to thrive as well!

Antiparasitics

There are many, many proclaimed dewormers. However, for this publication, only ingredients with historical or scientific support are listed. On pages 39–72 is that list with common names of the most readily available antiparasitics, and potentially the easiest to grow or obtain. If a scientific study for an ingredient was discovered, an asterisk (*) is placed after the name; otherwise, historical information is listed. Following the name will be a notation whether the compound acts as a vermicide (VC, killing the worm), or vermifuge (VF, destroying or expelling the worm), if the information was available. Next listed are the targeted parasites (P:) and/or its specific host, the dosage, and finally, other pertinent information including the parts of the plants utilized for the treatment. Before administering any of these remedies, please be aware of a plant's toxicity, and please do further research to determine if an ingredient will work appropriately for your needs.

Research into historical usage (pre-1930s) of several of the listed antiparasitics did not indicate the parasite the herb targeted. This could be for two reasons: 1) Scientific studies on specific species was limited, so when a traditional method was passed on, "Stomach Worms" or "Intestinal Worms" meant just about anything residing in a general area of its host. 2) Many antiparasitics do target several species of parasites, and not until actual studies were conducted did people discover *which* dewormers worked best on which parasite. Additionally, scientific research for herbal remedies is very limited, especially in the United States. Our societal addiction to conventional, chemical treatments is deeply rooted, and has been since around the 1940s. So, if a specific parasite was not cited in the research for the use of an ingredient, "GEN" will mean "general antiparasitic."

Lastly, most herbal treatments work best when in liquid extract form, since the herb itself (or dried matter) will not have to go through breakdown/digestion, and then absorption. Basically, liquid extracts go almost immediately into absorption. Some folks prefer the more "natural" approach of offering fresh ingredients as much as possible, while others appreciate the more concentrated effort of extracts given as a drench. The ingredients list does not have all methods given, so use your best dosage estimate if you choose a different manner of administering the ingredient. Treating animals on an empty stomach is recommended. If possible, administer antiparasitics at least one hour before animals are fed.

Special Note: At the time or writing this, I could find *no* research or historical information about natural compounds that work against meningeal worms (*Parelaphostrongylus tenuis*) once they have invaded the spinal column. I will not recommend remedies, but the chemical antiparasitics Ivermectin and Fenbendazole (in much higher doses than normal) have been proven to work against this parasite. Using chemicals will affect milk in dairy animals and there are withholding periods before consuming this milk. If you resort to chemicals, it may be best to isolate the treated animals (sacrifice pen) to keep the parasite area contained and manageable. Do further research to determine the right course of treatment for your livestock/system(s).

List of Antiparasitics
KEY:
* = Scientific study found   **VC** = vermicide **VF** = vermifuge **P** = targeted parasites
GEN = general antiparasitic **BW** = body weight **Unk.** = Unknown
Dosage = amount of material per pound(s) of animal's body weight, listed in standard (and metric)

Aloe*—P: Eimeria (coccidiosis); studies conducted with chickens and sheep. Dosage: 1 tsp. per 10 lb BW (4 g per 4.5 kg BW). Although dose duration could not be established, the Eimeria life-cycle would dictate aloe be administered every 14 days until symptoms subside. With conventional treatment, many lambs and kids are started on coccidiosis prevention beginning at 6 weeks of age and are given medication at least twice with 30 days in between. Piglets are started as early as 2 weeks old. Some research states to

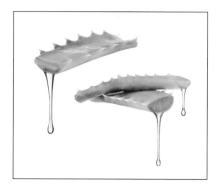

Aloe goo

only treat "infected animals," and to avoid "medicating" animals who don't present symptoms. The plant's succulent leaves are smashed and the gooey "syrup" used in treatments. In cool storage, aloe syrup will keep up to a week.

Berberines*—P: Giardia; study on children. Dosage: ½ tsp. per 100 lb BW (3 g per 45.4 kg BW) per day for 10 days. Berberine is found in the leaves and roots of goldenseal, barberry, Oregon grape, and yellow root. Collect and dry the herbs, then crush into powder for use.

Black Walnut* (VC)—P: Pinworms, roundworms, tapeworms. Dosage: ⅛–¼ tsp. per 100 lb BW (½–1 g per 45.5 kg BW) twice daily for seven consecutive days. Juglone is the active ingredient in the green hull which must be collected from the tree and dried from fresh material. Do not use if found on the ground. Strip the green hull from the walnut and place in a warm, dry place. If using a solar dehydrator, keep the hulls

Various stages of black walnuts

out of sunlight. Depending on the heat levels, expect the hulls to be dry in 3–7 days (when dry, they should snap when bent). Crush into a powder. (Do *not* administer to horses or pregnant animals.)

Brassicas—P: GEN. Most commonly used: white, black, and yellow mustard seed. Dosage: 1 tsp. per 100 lb BW (4 g per 45.5 kg BW) of dried material, or 4 oz/100 lb BW (120 mL per 45.5 kg BW) daily, of fresh material, such as radish, turnip, rutabaga, and horseradish. Kale and collards are in this family as well, and are great to feed as a rich source of calcium and other minerals.

Carrot Seed (wild, Queen Anne's lace, or cultivated)—P: Lungworm, barber pole worm. Dosage: (Unk.). Member of the Umbelliferae family, use seed.

Cayenne—P: GEN. The active ingredient in the oil, capsaicin, helps kill parasites, especially when used in combination with other ingredients. Dosage: ½ tsp. per 100 lb BW (2 g per 45.5 kg BW).

Chrysanthemum—P: Ascaris, strongyles, and all external parasites. GEN. This flower produces the insecticide pyrethrum. The powder from the dried and crushed plant material can be added to feed or deworming treats. But, like tansy and other colorful antiparasitic flowers, Chrysanthemum daisy is a wonderful addition to beneficial plants areas. Internal dosage: Just a pinch (less than ⅛ tsp) per 100 lb BW (157 mg per 45 kg BW).

Cloves—P: GEN. Dosage: 3 cloves, crushed into powder) per 100 lb BW, (approx. 25 mg per 45 kg BW) 2x day for 7–10 days.

Conifers—(Also called evergreens. Includes bald cypress, pine, cedar, and fir trees) P: Flukes, GEN. Dosage: offer approx. 2 lb fresh material per 100 lb BW every day for 10–14 days. Note: turpentine (a terpene) can be extracted from conifers' sap and processed into an anthelminthic. However, the sap from these trees is quite volatile, and the process of extracting, storing, and using turpentine will not be provided in this booklet. However, give livestock free access to nipping at the bark of these trees during high-load times.

Copper* (VF)—P: GEN. For goats and cattle only. Copper is highly toxic to sheep and horses. Dosage: Copper sulfate treatment used during high-load times is ½ tsp. per 100 lb daily for up to three weeks. Copper sulfate can be offered freely as a mineral lick at 1 lb copper sulfate, 1 lb sulfur powder, and 1 lb seaweed or kelp. Lastly, as a drench mix: 5 ounces of copper sulfate with 1 gallon of water then administer 1cc for every 5 lb of body weight. Copper wire particles administered at the rate of 12 g per 500 lb BW (or 4 g per 125 lb) via a bolus can also be administered in early spring, then again in middle of fall (during wet, high-load times).

Cucurbits (VC)—P: Barber pole worm, liver fluke, pinworm, roundworm, tapeworm. Dosage of *C. maxima*: 20 seeds per 100 lb BW daily for 2 weeks. Increase the dosage 10 additional seeds for *C. moschata*, and 20 additional for *C. pepo* species of squash. Lightly crack/crush for poultry, crush into powder for mixing as a drench, or leave whole for cattle, horses, goats, sheep, and pigs and add to feed.

There are hundreds of varieties of squash, and their seeds vary in size and color.

Echinacea purpurea—**P:** GEN. Contains tannins and is widely used for immune support. Dosage: ½–1 tsp. per 100 lb BW (2 g per 45 kg BW). It was tested in several large dairy goat herds in New Zealand where a teaspoon of ground dried root material per goat, added to feed, controlled internal nematodes. No indication of duration, but according to general use precautions, treatment should not exceed 14 days, then repeat (if necessary) two weeks later. Grind dried herb and/or root into powder for use, or offer in beneficial herb paddock.

Fennel*—P: Barber pole worm, threadworm. Dosage: (Unk.) In Russia, sheep infected with barber pole worms were grazed on fennel-covered pasture for up to 20 days, although fecal counts showed the worms were eliminated after only two days. (Fennel is a member of the Umbelliferae family, like carrot, and the seed too can be used.) Dry the leaf and crush into a powder, and add fennel to the beneficial plants area.

Fern (Marginal Wood—or Shield—Fern) (VF)—P: Tapeworm, but *not* flukes or echinococcus of dogs. Dosage: Using young shoots (fiddleheads) and rhizomes ½ tsp. per 100 lb BW (2 g per 45 kg BW), **one time only**. HIGHLY TOXIC TO LIVESTOCK, use with caution. Dry and crush the material for use.

Freshly picked fern fiddleheads

Fineleaf Fumitory/*Fumaria parviflora**—P: Barber pole worm, brown stomach worm, most roundworms, small intestine worm, nodular worm, whipworm. Dosage: The freeze-dried "Aqueous Liquid Extract (ALE)" used in the study is described at the end of this "Ingredients" section. Researchers used approximately ⅛ tsp. per 10 lb BW (0.5 g per 4.5 kg BW). This plant should not be confused with *Fumaria officinalis*, which does not have anthelmintic properties. Fineleaf Fumatory is found primarily in the Southwest.

Garlic* (VC)—P: Ascaris, lungworm, giardia. Dosage: 1 med. clove per 50 lb BW. For poultry: Crushed cloves, 1 clove/5 chickens, can be soaked overnight in water which was brought to a boil, then added to poultry's drinking water with a 1:4 parts ratio, or 1 cup garlic water to 4 cups drinking water. Some poultry and other livestock will eat crushed garlic added to their feed. If using dehydrated, powdered garlic, use about ⅛–¼ tsp. as a 1 clove equivalent. Garlic prevents eggs from developing into parasites, but does not kill the actual parasite. Garlic is usually used in conjunction with other herbs.

Ginger*—P: Roundworms, flukes, heartworm in dogs and cats. Dosage: 1 T. per 10 lb BW (12 g per 4.5 kg BW) X 10 days (fresh herb). If dried and shredded, use ½ the amount, and if dried/powdered use ¼ the amount. Note that heartworm is a life-threatening condition for dogs and cats. Readers should certainly seek veterinary assistance rather than relying on home care alone.

Goosefoot/Pigweed—P: GEN. Feed the plant directly to animals or grow it in a beneficial plants paddock. Tea made from the leaves can be used as a drench, and the dried and powdered seeds can be added to deworming treats. Dosage information unavailable.

Grapefruit Seed: Dried/Powdered or liquid extract—P: All parasites, giardia. Dosage: Extract (per 100 lb BW); 1–2 drops for two days, 2–4 drops for three days, 4–6 drops for two days. Powdered/Capsule; 200 mg per day for 7 days. Both extract and capsules are usually purchased from retailers. It is possible to grow and produce your own; however, the process will not be covered here.

Indian Ginger/Snakeroot*—Studies done with cattle and horses. Dosage: Root; 1 tsp. per 100 lb (4 g per 45 kg BW). Be careful how much is harvested, as this herb is slow to regrow.

Juniper Berries—P: Flukes, GEN. Also called "savin," these berries are in the conifer group. Dosage: (Unk). Offer free choice in the fall when berries are ripe. Do *not* use with pregnant animals.

Lupine/Lupins—P: Trichuris, Strongyloides, Ascaris, Strongylus. Mix into the beneficial plants paddock. Make sure there are many other plants from which the animal can choose so they are *not* limited to eating only lupine. Too much will cause poisoning in all animals. Specific dosage unavailable.

Nasturtium—P: GEN. There are two "nasturtium" plants in popular research. Both are edible, and both have anthelmintic properties. Watercress, or *Nasturtium officinale*, is more "toxic" than the garden-variety nasturtium, or Tropaeolum. The garden nasturtium gets its name from the fact that it produces an oil similar to that produced by watercress. In either case, dosage information is not available, but the garden nasturtium should be a beneficial plant used all around livestock pastures, including poultry, to be eaten as needed. NOTE: Watercress should not be ingested by pregnant animals.

Orange Oil Extract*(VF/VC)—P: Barber pole worm, GEN. Dosage: 1 tsp. per 10 lb BW (5 mL per 4.5 kg BW) for 2 days. Orange oil can also be used as a broad-spectrum spray in pastures to kill parasite larvae. However, orange oil will also kill other, beneficial larvae and insects, so if possible, spray only in a 12-inch circle around dung heaps, or in small, paddock-type confinement areas. Orange oil is volatile and MUST be mixed and administered with a vegetable oil in an approximately 50/50 ratio. Example, if your lamb weighs 40 lb, it would receive 4 tsp. of orange oil mixed with 4 tsp. of vegetable oil.

Oregano (VF)—P: GEN. Contains thymol as the active anthelmintic. Dosage: (Unk.). Another nontoxic plant to add to a beneficial plants pasture/paddock and allow for choice-feeding of all livestock types. Can also be made into an aqueous solution for drenches, or offered freely in feed troughs.

Rosemary (VF)—P: GEN. Dosage: 1 tsp (4 g) per 100 lb BW (4 g per 45.5 kg BW) per day. Rosemary is in the mint family, and as other mints, possesses anthelmintic properties. Another beneficial herb to add to the pasture or offered freely in feed troughs during high-load times, or made into a drenching solution.

Sage* (VF)—P: GEN. Most of my livestock did not eat any kind of sage, but it's an easy addition to a beneficial plants paddock.

Tannins*—P: Barber pole worm, giardia, GEN. Tannins are found in the leaves of heather (19% tannins), chestnut trees (24.7% tannins), hazelnut trees (14.2% tannins), oak trees (5.3% tannins), and *Sericea lespedeza* (avg. 11% tannins) to name a few. Studies across the board indicate feeding *Sericea lespedeza* (SL) hay or high-level tannin leaves reduces the parasite load in goats and sheep from 60 percent to 80 percent. However, studies also show that concentrated feeding (over 75% of the total ration) of high-concentrate SL has resulted in death among cattle. As a guide, then, plants with the highest percentage of tannins should never exceed 75 percent of a ration with any livestock. It would be best to graze animals on pastures with high-tannin plants (listed above) during high-load times.

Tansy (VC)—P: GEN. Large roundworm. Dosage: All plant parts except the root; soak overnight 2 T. in 1 pint of water (30 g per ½ l. water) that has been boiled. Use 2 oz (60 mL) as a drench 2X/day until the liquid is gone, or for a larger batch, soak overnight 4½ c. of chopped leaves, stems and flowers in 4½ c. of water (1 kg of material per 1 l. water) that has been boiled. Then use ½ tsp. per 10 lb BW .5 mL per kg BW) 2X/day for two doses, one day apart. Do *not* use with pregnant animals. Tansy contains thujone. One source indicated that goats, horses, and pigs should not receive tansy.

Tarragon*—(Artemesia family) P: GEN, roundworms. Only listed in research as a bioactive forage, used in beneficial plants pasture/paddock for free-choice grazing.

Thyme (VC)—P: GEN. This herb is in the Artemisia family, like wormwood. It has been described as a beneficial plant to be free-grazed. Do *not* use with pregnant animals.

Tobacco—This highly toxic plant is most successful with poultry. In the 1920s the USDA recommended this plant as a dewormer. To date there are no available dosage amounts to be found. HIGHLY TOXIC, use with caution.

Turpentine—P: Ruminant liver flukes, horse strongyles, GEN. Turpentine spirits are a by-product of distilling turpentine. Dosage: Pure turpentine spirits: 75 drops per 100 lb mixed with two ounces of food-grade vegetable oil or seed oil. Add rolled oats or other crushed grain and/or molasses, with enough to mask the odor. This old-time cure was used frequently pre-1920, and administered via "tubing." However, it is as potentially harmful as tobacco. USE WITH CAUTION: If turpentine enters the respiratory tract, involuntary closure of the tract may occur!

Wormwood* (VC)—P: Most all helminthes except tapeworms. Studies have shown coccidia can also be reduced with its use. Wormwood is in the Artemisia (A.) family with over 300 species, including sage, mugwort, tarragon, and obviously, the many wormwood varieties, which makes it an easy addition to your herb garden or beneficial plants pasture. In research, *A. absinthium* (grand wormwood) repeatedly performs better against more species of parasites with higher VC/VF rates than other Artemisias. Next is *A. vulgaris* (mugwort), *A. herba alba* (white wormwood), and *A. annua* (sweet Annie), all with great antiparasitic properties. Because the efficacy varies with each variety of wormwood, there were several dosage levels tested. All methods of administering the herb were effective. Dosage: 1) Oral dose of dried herb 2 tsp. per 10 lb BW (1.5 g per kg BW), or 2 T. (20 g) per 100 lb BW, or, as an 2) Aqueous solution: 2 T. per 16 oz (25 g of material per 500 mL of water) that has been boiled and has set for 72 hours then been filtered. Either should be administered in the morning on an empty stomach. There was no indication in research for the number of days to administer wormwood, but several human-dosage studies indicated from three to seven consecutive days. Please do more research to choose the best method for your needs. Note: Do *not* use any wormwood or members of the Artemisia family on pregnant animals.

Yarrow* P: GEN. Dosage of dried material: 1¼ tsp. per 10 lb BW (6 g per 4.5 kg BW) *one time only*. Yarrow grows and propagates easily in the wild. Found in fields and pastures almost everywhere. Yarrow can be confused with poison hemlock (which is very toxic) during flowering, and with Queen Anne's Lace before fully flowering. If you are collecting this plant from the wild, do further research to ensure correct plant identification. **NOTE:** Yarrow is poisonous to calves.

As a general rule of thumb:
- Garlic and conifers are good as preventive dewormers for regular use. (Use in lactating dairy animals may result in off-flavored milk, though.)
- Aromatic herbs like fennel, oregano, rosemary, sage, tarragon, and thyme can be offered freely, and/or in a beneficial plants paddock. Up to approximately 1 T. per 100 lb can be given once a day for up to seven days. Many animals do not like these strong herbs added to their regular feed, so offer them in a separate feeder.
- More potent natural antiparasitics include wormwoods, Indian ginger (snakeroot), goosefoot (pigweed), cucurbits, Umbelliferae species, black walnut, and

tansy. These either kill or paralyze parasites so they release their hold in the digestive tract and are expelled. Some of these possess abortive properties, so use with caution.

- Fern, turpentine, and tobacco are only administered *one time* and have serious side-effects. Only use these when other remedies are unavailable or not working.

Dosages: Unless otherwise indicated for an individual ingredient, *recommended dosages should be given for three to seven consecutive days.* Administering at least one hour prior to feeding is most effective. And, when animals are housed and/or fed repeatedly and consistently in the same area(s), deworming should be done *every 30 days.*

* For best results, do all three methods simultaneously: 1) deworm, then 2) move livestock to a fresh pasture, and 3) clean and treat housing, and/ or move and treat around round bales and feeders.

Besides using dried plant material, here are other ways to prepare antiparasitics:

Aqueous Liquid Extract (ALE) process: Boil dried plant material for 1.5 hours. Cool to around 70°F. Strain through fine mesh or paper filter. Place liquid into low, saucer-type dish and let the water evaporate. Gently scrape the residue and store.

Aqueous Solutions/Drench: Bring water to a boil, remove from heat. Steep the herb(s) (either fresh or dried) from 4 to 12 hours. Strain the leafy material out and use the "tea" as a drench. (I use about ½ c. water per 100 lb.)

Molasses: For most hoofed livestock, including pigs, adding molasses to dried plant material makes it more appealing. Also, hand-fed treats or "cookie-balls" made ahead of time work well with tame livestock.

Bolus: Some of the more potent ingredients can be made into boluses (various livestock sizes ordered from a veterinary supply), such as the black walnut hull and chili pepper . . . *if* you are skilled at administering boluses.

Chart D (pages 78–79) lists the most common parasites, including their common name, scientific classification, the areas the parasites affect, and the treatment ingredients used. The scientific classification information is added for one specific reason: it is reasonable to surmise that the treatments for a certain species of parasite may well work against other species within the same group. And, if you cannot get a certain herb, try one of the ingredients prescribed for a parasite in the same class (and better yet, same genus). Unfortunately, I was unable to find a single scientific study confirming the use of an herbal treatment that worked on all species within a specific group. But it's still the best way to choose a substitute herb.

If you are not sure which parasite is afflicting the animal and cannot have a fecal sample tested, it would be wise to choose at least three ingredients to administer jointly. This can be done as a drench in combination with apple cider vinegar or boiled water. Or, choose a few fresh herbs from the list to offer in a feed trough so the animal can pick the one(s) it needs. Some folks even make "treats" with oats or ground corn, molasses (or another sweetener like stevia), and the herbal remedies rolled up all together. Choose methods that work best for you and your animals.

And, remember to do all three practices simultaneously: 1) deworm (empty stomach is best!), 2) move livestock to a fresh pasture, and 3) clean and treat housing, and/or *move* and treat around round bales and feeders.

Following are a few suggested combinations; these do not have the dosage amounts, because the animal body weight determines the dose. When creating your recipe(s) remember this: Each and every ingredient in your concoction must be the recommended amount. For example, if using wormwood, garlic, and mustard seed, the *full* dose amount of *each* (as determined by body weight) is combined together for a treatment.

Example Ingredient Combinations

Cats and Dogs: Crushed cucurbit seed, wormwood, black walnut. Add the recommended dosages amounts to wet pet food.

Cattle: 1) Cucurbit seeds, 2) Mustard seed, wormwood, berberines. Dairy cows will sometimes eat deworming treats made with cornmeal and molasses, otherwise add to feed or grain.

Fowl: 1) Crushed cucurbits, grapefruit seed, mustard seed. 2) Garlic, mustard seed, nasturtium, tansy, wormwood. Mix dosages into moistened grains.

Goats or sheep: 1) For pregnant animals: Cucurbits (crushed or whole), garlic, ground mustard seed, Oregon grape root powder. 2) Nonpregnant animals: Cayenne, yellow mustard, garlic, black walnut hull, wormwood. These ingredients can be steeped together to create a drench, or made into deworming treats by mixing all ingredients with molasses and forming into balls. Refrigerate to store and use for the dosage duration. 3) Free-choice dewormer lick: ½ cup each cayenne, mustard, garlic powder, and a natural trace mineral.

Horses: Cucurbits, garlic, ginger, sage, wormwood. Use the recommended dosage of dried herbs and crush into a powder. Add water or ACV to make a thick paste. Some people add up to a tablespoon of molasses as well. Using a drenching syringe, administer your mixed paste as you would any conventional dewormer. Horses can also be fed molasses treats.

Pigs: Cucurbits, garlic, tansy, wormwood. Pigs will eat just about anything when it's added to their food.

Rabbits: Cucurbits fed whole with feed at the recommended dose. Offer fresh herbs, or place rabbits in moveable tractor over areas with beneficial plants.

Example Treatment Calculations

First, determine the parasite(s) affecting the livestock either by conducting a fecal check or having it done by a veterinarian. Next, study Chart D (pages 78–79) to determine the ingredients you will need to eliminate those parasites. Finally, peruse the "List of Antiparasitics" (pages 39–72), and determine the dose amount of the ingredients and its corresponding animal's body weight. Now, it's time to put your math skills to work.

When writing this section I tried to get family and friends to read and use the methods I've described. One of those people has seven hens, so I sent him the instructions. Here is an excerpt from his email: "*My brain imploded! There was a lot of addition, subtraction, multiplication . . . calculations of all sorts. It would be WAY easier if it could just be addition. I found it easier just to mix it (one day at a time, every morning) all three days in an 'addition only method.'*"

This *is* one way to do it. Using the "List of Antiparasitics" (pages 39–72) anyone can just "add up" from the dosage listed. However, most of those dosages are for 100 pounds of body weight. If someone has only four chickens, or two turkeys, or three barn cats, there will still be more math involved. But, it might still be manageable for six mornings in a row. However, if you have more than a few animals, counting out multiple ¼ teaspoons of many ingredients every morning for six days *will* become a tedious, full-time job. I have used two ways to calculate how much of each ingredient I will need for three days of dosing and, for another three days of free-choice lick (goats, sheep, cattle, etc.). Before I break it down, here is a short list of necessary measurement conversions:

3 teaspoons (tsp.) = 1 Tablespoon (T.)
4 T. = *¼ cup* (c.)
1 ounce (oz)/6 tsp = 30 cc (cubic centimeters)

Example 1: Jane Doe's 14 dairy goats are kept in a dry lot from December until the end of February. They are pregnant and cannot have certain antiparasitics, like wormwood or black walnut hull. She creates molasses treat balls for 10 older does, and makes a drench for four young does who won't eat the treats.

The first method works by starting from the smallest amount and multiplying up to the total amount. Here's how she does it:

1. She estimates the goats' weights. She finds that most are around 100 lb. each.
2. Jane chooses from the "List of Antiparasitics" (pages 39–72) plants she can easily grow or buy organic online. She uses the ingredient amount per body weight from the dosages in the list.
   a. Berberine; barberry (½ tsp per 100 lb)
   b. Brassica; yellow mustard seed/crushed (1 tsp per 100 lb)
   c. *C. maxima* squash seeds (20 seeds per 100 lb, or about 1½ T. crushed)
   d. Cayenne pepper (½ tsp. per 100 lb)
   e. Garlic powder (¼ tsp. per 100 lb)
   NOTE: The total dry matter is about 6¼ tsp. per 100 lb, or 2 slightly heaping T. of mix per goat per day, and 6 slightly heaping T. for three days.
3. She calculates the total amount of each ingredient she will use for six days by multiplying:
   (Ingredient) x (number of goats) x (number of days)
   <u>Jane's Recipe</u>
   a. Barberry: ½ tsp. x 14 goats x 6 days = 42 tsp. (or 14 T., or ¾ cup + 2 T.)
   b. Yellow mustard seed/crushed: 1 tsp x 14 goats x 6 days = 84 tsp. (or 28 T., or 1½ cups)
   c. Crushed *C. maxima* squash seeds: 1½ T. x 14 goats x 6 days = 126 T. (or about 2 cups)

d. Cayenne: ½ tsp. x 14 goats x 6 days = 42 tsp. (or ¾ c.+ 2T.)

e. Garlic powder: ¼ tsp. x 14 goats x 6 days = 21 tsp. (or 7 T.)

4. She weighs this total amount and splits it into two equal parts. Half will be for the three days of manually administered treatment, the other half for free-choice lick.

5. Now, remember how much one dose was (note from #2 above). In one bowl, Jane puts enough dry matter for the four who will be drenched for three days: 2 T./day X 4 goats X 3 days = 24 T., or 1½ cups. (She sets aside the remaining mixture for the 10 who will be getting molasses treats for three days.)

6. For the drench: she boils about ½ c. water or apple cider vinegar per animal's total three-day dose, or 2 cups. She removes it from the heat and stirs in the material to be made into a drench. She covers and lets it steep for at least 4 hours (12 hours is preferable).

   a. Jane thoroughly presses all liquid from the ingredients. (You can save the wet material for chickens, pigs, dogs, etc. Although it's at about ¼ of its original potency, it's enough to treat a 100-lb pig once!)

   b. She adds a couple tablespoons of molasses (which makes the drench taste better for the goats), and divides this liquid into three equal parts (one for each day) into containers with measurement capabilities (beaker, canning jar, etc.).

   c. Then, she divides each daily amount into four equal servings for each goat who will be drenched. (If using a wide-mouth pint jar with four ounces, I know each goat will get one ounce per day.)

7. For the treats, Jane takes the remaining dry material and adds molasses a bit at a time until the mixture resembles cookie dough.

   a. She weighs the total amount, then equally divides that into three parts, one part for each day.

   b. For each day's batch, she weighs and scoops up 1/10 of the "dough" and rolls it into a ball. She puts 10 treat balls per day into a container and refrigerates.

8. One hour before feeding, Jane can drench and/or feed treat balls to the goats.

9. After three days of drench and/or treats, Jane will use the remaining dry mixture to make up a free-choice lick. She also adds ¼ cup of sugar and ¼ cup of goat mineral to the total mix (some folks also add kelp and/or baking soda). Then, she divides it into three days and offers the mixture in a free-choice trough.

Example 2: The second method of calculation starts with using the animals' total weight. In the above example, Jane's total was 8,400 lb (14 goats x 100 lb. each x 6 days). Using the formula below, *for each ingredient* you will do the following;

1. Write a single dose amount on the top line in column (a)*, and the body weight for that dose on the bottom of column (a) #.
2. Write the total pounds on the bottom of column (b) ##.
3. Multiply the top * (a) with the bottom number ## in (b).
4. Divide *that* total by the bottom number # in column (a) to arrive at the total ingredient amount (column b, top).

	(a)	(b)	
ingredient	___*___	(_____)	-total ingredient
BW/pounds	#	# #	-total pounds

Here's an example for the cucurbits at 1.5 T. per 100 lb:

	(a)	(b)
amount	1.5 T.	(126 T.)
pounds	100	8,400

Multiply 1.5 X 8,400 = 12,600, divided by 100, or 126 T. (You can see in number 3c above, the end result is the same.) Here is the equation for using garlic in this specific recipe.

	(a)	(b)
amount	.25 tsp	(21 tsp.)
pounds	100	8,400

Multiply .25 X 8,400 = 2,100 divided by 100 = 21 tsp. (#3-e).

Using this calculation method I've also created "bulk" mixes so monthly deworming takes less time. After measuring out a single dose of all ingredients I'm using, I write that on a sticker (example: 2 T. + 1 tsp.) and put it on a large container. I then combine the larger amounts of the ingredients (using the math formula above) for up to100 doses (or more) and mix well. Then, I can get any number of doses anytime from the bulk container and mix with feed, use as a drench, or prepare bulk boluses!

Here's one more calculation that's helpful for those of us with chickens. It's a recipe I use for six consecutive days, based on the average chicken weighing about 10 lb. Many layers are usually 7–8 lb, but using 10 pounds makes for easier math, and the little bit of extra wormwood will not kill them . . . only the parasites! I mix this into moistened morning feed and keep them confined for one hour before letting them free-range. For every 10 chickens I use:

a. Barberry: ½ tsp. X 6 days = 3 tsp.
b. Yellow mustard seed/crushed: 1 tsp. X 6 days = 6 tsp.
c. Wormwood: 1.5 T. X 6 days = 9 T.
d. Cayenne pepper: ½ tsp. X 6 days = 3 tsp.
e. Garlic powder ¼ tsp. X 6 days = 1½ tsp.

Weigh the total, divide that by six (days), then add that 1/6 portion to the morning moistened feed.

I truly understand how this part of the plan can feel a bit hard to grasp and to carry out, and the math can feel cumbersome. I believe that's the number one reason folks fall back on chemical treatments—they're easier, faster, and already measured out. But they're also a *lot* more expensive. Studies on natural anthelmintics are most often conducted in third world countries where farmers can't afford a bottle at $160! If you do find yourself having to fall back on conventional methods, try to keep the animals in a dry lot or use a sacrifice pasture during chemical use. This will reduce the impact of those chemicals on the environment.

Growing, Collecting, Processing, and/ or Storing Antiparasitics

One of the problems with trying to suggest growing procedures is that nothing starts or grows the same for any two people because of zones, soils, specific property characteristics, and/or an individual's care of seed(ling)s and plants.

For instance, I could never get echinacea (purple coneflower) to start from seed. I tried for several years, then finally I bought a dozen young plants, researched where they'd do well on my property, planted them, and they took off! I did the same after several attempts to start chrysanthemums (pyrethrum). After researching their needs, I ordered the plants from a reputable online nursery and put them in the ground. Again, success! Yarrow (usually highly invasive) was also hard to start and, even though I did get some plants to start after cold-stratification in the freezer, only a few of these "weak" plants returned the following year. I also sowed yarrow seed in the ground in the fall (as suggested on the package), but only a couple seedlings came up in the subsequent spring. Then, after research, I decided to wait until later in the winter to sow it (to reduce insect and rodent consumption, and to stop the chance of rot from my area's fall rains). I sowed it into the ground in January . . . and, voilà! It too took off. Only after you thoroughly research the plants' needs of location, soil, and water (and don't just read the instructions on the seed packet) will plants thrive.

Understand that there are many, many variables, and there really needs to be "the perfect storm" (sometimes literally) of these variables for plants to get established. It

Herbs drying

can take several years' attempts. But, that said, here are the rules of engagement when trying to develop the elusive green thumb: First and foremost, be sure that you are working with plants recommended for your USDA Hardiness Zone. Next, research the sunlight (full, partial, shade), soil, and water needs. If moist soil is needed, make sure your plants are near a water source. And, if they need a lot of water to start, make sure this will be easy to do with your schedule. Watering baby plants two football fields away from the spigot, especially during kidding season or when putting in your garden, might be a little tough to accomplish. Also, get soil samples of prospective areas to ascertain the pH levels, and research the type of soils these selected plants need to thrive (sandy, mulched, clay, etc.). Remember, it's better to keep plants in an area that will not need constant pH adds or modifications. Simplicity is best; the less you have to modify the better. And sometimes using native or "wild" versions of plants (not the cultivated species) can also prove to be very successful . . . if you know these plants' requirements.

Below is a list of anthelmintics' basic growing and harvesting requirements. Remember, do specific research on the one(s) you choose to grow and harvest to ensure your chosen growing plots will be successful.

Aloe (zones 9–11): In cold climates, aloe plants must be kept from freezing. They propagate themselves, and quickly! I've always kept mine in pots so I can bring them inside in the winter. To collect the "sap," cut thicker stalks in half and scrape out the gooey middle. Keep refrigerated and use within 3 days.

Aloe

Berberines: Goldenseal (zones 3–8), barberry (zones 4–8), Oregon grape (zones 5–9), yellow root (zones 4–9). Barberry is by far the most practical to harvest and use, as it grows quickly and is invasive in almost everyplace it grows. Yellow root too is a fast-grower, and is a great addition to a semi-shady plot of land for you to grow and harvest responsibly. But, goldenseal and Oregon grape have been consistently overharvested because of their medicinal benefits. These plants are slow-growing and use of them in anti-parasitic mixes should be limited (and only purchased from a responsible, reputable retailer). The root is the material used. Roots are tough and always difficult to process! Collect the root from three-year-old plants or older. Wash thoroughly, pat dry, then chop into smaller pieces. Dry/dehydrate. Grind into powder.

Goldenseal

Oregon Grape

Barberry flowers

Barberry berries

Black walnut (zones 5–9): These must be collected when the hulls are green, otherwise, as the hull turns black lying on the ground, the juglone becomes inert. After collecting them, peel the green hull from the nut, throwing away any blackened parts. Dry the hulls in a well-ventilated, hot area, out of the sun. If using a solar dehydrator, cover the green hulls with a light towel to keep them out of sunlight. Depending on the heat levels, expect the hulls to be dry in

Black walnut with green hull covering the nut

3–7 days—they should snap when bent. Crush into a powder. (*Do not administer to horses. Do not use in pregnant animals.*)

Brassicas: Brassicas include kale, broccoli, cabbage, collards greens, and white, yellow, or black mustard seed and nasturtium seed. These grow best during the cool season, thriving from fall through spring in zones 9 and 10 and growing as spring or fall plants in colder areas. I've always ordered my mustard powder, because the seeds are so small and it takes a lot of plants and space to grow enough to make your own. Nasturtium seeds are larger and the flowers are great in the garden (as a beneficial insect attractant) and all over the farm where all livestock can browse as needed. Additionally, I feed cabbage, kale, and broccoli leaves and stalks all through the summer as I trim "bad" stuff from the garden.

Wild yellow mustard

Carrot seed: Queen Anne's lace (wild carrot, zones 3–9) is wonderful in the pasture if you have goats or sheep who seem to ravenously snip the flowering heads from their stems at peak season. As well as letting livestock eat the flowers and seeds, their roots can be harvested and fed fresh and/or dried. But Queen Anne's lace root can be quite tough! Chop it into small pieces to dry more quickly, then grind

Queen Anne's lace

it into a coarse powder. Queen Anne's lace can do well in even the toughest of soils as long as you can get it started. Conversely, cultivated carrot (zones 3–10) love sandy, loose soil and cool temperatures. It requires two years before it will go to seed. It's best to cover the chosen stock of carrot root with a layer of fluffy mulch so it doesn't turn to mush during the winter freeze and subsequent spring thaw. Better yet, let Queen Anne's lace do the work! To ensure some of the plants go to seed, don't let livestock eat the last of the summer's existing plants. (Especially before flowering, Queen Anne's lace can be confused with poison hemlock. Please ensure you have properly identified the plant before using it, as poison hemlock is deadly.)

Cayenne (zones 9–11): Cayenne is one of the easiest plants to grow in your garden. When the majority of peppers have turned bright red, pluck the entire plant, then hang it upside down in the garage until the peppers are all dry. I pick the peppers from the dried plant, remove the stems and any rotten or molded ones, then crush the entire pepper into red flakes. I wear standard rubber household gloves while working with cayenne, as the active ingredient, capsaicin, will cause a burning sensation on the skin.

Cayenne peppers on the plant, with a few blossoms

Chrysanthemum (Pyrethrum) (zones 3–9): These seeds seem to do better if they are put in the ground in fall and can have a winter's worth of freezes (cold stratification). You can also order small plants from a nursery and plant them in spring, allowing at least 50 percent of the flowers to go to seed and replant themselves. These plants prefer full sun and ordinary soil. Cover lightly with mulch in the fall and many root balls will regrow in the spring. Harvest the flowers throughout the year and dry them. After being dried, the flowers will contain approximately 30% pyrethrum.

This variety of chrysanthemum is also called the pyrethrum daisy.

Cloves: Cloves will only grow in subtropical and tropical climates, never below 50°F. I've always ordered organic crushed cloves. Additionally, greenhouse growing "enough" clove for livestock would be space-consuming.

Conifers (also called evergreens. Includes bald cypress, pine, cedar/juniper and fir trees): I have so many wild pine and cedar trees, I allow the animals to eat what they need. I have not collected, dried, or processed these materials. I do know that my goats *love* the bark of cedars, and strip baby trees until they're dead. So, if you want cedars to grow, keep your goats away from them!

Juniper berries are in the conifer family, and goats love them!

Copper: I've ordered copper sulfate and have sprinkled the dosage amount on feed with molasses and added it to molasses treats as well. I've also given copper oxide wire particles (COWP) to my goats and administered as an oral bolus, which is time-dissolved over the course of 3–4 months.

Cucurbits (squash) (zones 3–10): In my garden, I've grown one from each family: maxima, mixta, moschata, and pepo. Keeping only one variety will stop cross-pollination so you can use seeds saved from a previous year. When squash is mature, I slice it in half, scoop out the seeds, remove the slimy stuff, rinse them, then dry them on plates on top of my cabinets and/or refrigerator for about 2 weeks. The "meat" of the squash is many times fed to the animals, and I've even began drying "chips" of squash to feed to livestock over the winter. I rely heavily on cucurbit seeds in the winter while animals are pregnant and cannot use wormwood or black walnut hull.

Echinacea purpurea (zones 3–9): This beautiful "wildflower" (purple coneflower) grows in harsh, dry soils. Make sure you only collect roots from three-year-old stands. Grind dried plants and/or root into powder for use, or grow in beneficial herb paddock.

Echinacea

Fennel (zones 3–10): The wild varieties tolerate the cold best, some down to –20°F. The delicate bulb of the cultivated varieties will not tolerate cold winters, even covered in mulch. My goats *love* this member of the Umbelliferae family, which is in the same family as carrot. Dry the leaves and stems and crush into a powder. Also, grow fennel in a beneficial plants area. My goats devour them! I have to be careful they don't eat them too close to the ground, especially in the summer when rain is scarce. Close-cropped plants will be too stressed and not return the following year. Also, let a stand of your fennel produce seed. You can section off an area with electric fencing and a small, solar charger.

Fennel

Fern (zones 4–8): Use young shoots (fiddleheads) and rhizomes. Dry and crush the material for use. *Highly toxic to livestock*, use only when other dewormers are unavailable. I have not used this plant and I don't know if livestock would choose to eat these if available to forage on their own.

Fern with fiddlehead

Fineleaf fumitory/*Fumaria parviflora*: This plant is found primarily in the Southwest, and grows in only seven states. I have not used it. If this is something that grows easily in your area, do more research before using.

Fumatory flowers

Garlic (zones 3–9): I have three beds of garlic . . . you can never have enough! I've dried and crushed it, and used it fresh. Each clove of garlic will grow another entire bulb! After harvesting, use smaller heads by breaking apart each clove and replanting it. I add garlic to the free-lick mixes, and to my chickens' water during high-load times (spring and fall for my area).

Wild Ginger

Ginger (zones 5–8): Cultivated varieties cannot withstand winter. Wild ginger can tolerate mild cold, but, I have not hunted for or used it. Large quantities of ginger are needed for using as antiparasitics, so growing for livestock use in a greenhouse would require a lot of space. As an experiment you can purchase a large root of ginger (3–4 inches long) from the grocery store, slice and dry it, then, grind it into a powder. One tablespoon is needed for every 10 pounds of livestock body weight. You would need about 10 T. for one treatment for one cow, or 30 T. for the 3-day treatment, and another 30 T. for the following 3-day "free-choice lick" treatment.

Goosefoot/Pigweed (zones 4–8): Also called wild spinach or spinach tree. This grows just about everywhere, but most of my animals ignore it; only my sheep seem to eat it.

Goosefoot

Indian ginger/Snakeroot (zones 4–9): As this is a highly toxic plant, make sure to limit its growth, and that livestock have plenty of other forage to eat. If used regularly, be careful how much is harvested, as this plant is slow to regrow.

Snakeroot

Texas bluebonnet/Lupine

Lupine/Lupins (also called Texas bluebonnet) (zones 4–8): This plant will make *lots* of seed and reseed easily. A beautiful, early bloomer, I grow this in my vegetable garden. I can then collect seed at the optimal time and sow seeds in fall in other parts of my farm. If you're mixing this into the beneficial plants paddock, make sure there are many other plants from which the animal can choose so they are not limited to eating only lupine, since too much will cause poisoning in all animals. Specific dosages are unavailable.

Nasturtium (zones 9–11): There are two "nasturtium" plants in popular research. Both are edible, and both have anthelmintic properties. Watercress (*Nasturtium officinale*) is more toxic than the garden-variety nasturtium, or Tropaeolum. The garden nasturtium gets its name from the fact that it produces an oil similar to that produced by

Nasturtium

watercress. In either case, dosage information is not available, but the garden nasturtium should be a beneficial plant used around all livestock pastures, including for poultry, to be eaten as needed. NOTE: Watercress should not be ingested by pregnant animals.

Orange Oil Extract: Because this extract can potentially be very harmful, I rarely use it. As well, extracting the oils requires specialized equipment which is expensive, and I did not purchase or attempt the procedure. Additionally, when making extracts, most times only one or two plant materials can ever be prepared with that specific equipment because of residues. Therefore, the process will not be covered here. Please do further research.

Oregano (zones 6–10): Easily grown and a perennial in areas where the winters are not too harsh. It may also survive zone 5 winters with a light mulch cover during winter. There are a few varieties that may tolerate even colder climates, so check with your county Extension office or USDA branch for recommendations. Cut, dry, then crush, and save in airtight container. You can also offer it fresh or grow in a beneficial plants paddock. It will not tolerate dry, hard soils, though.

Oregano

Rosemary (zones 5–9): Only a few varieties can tolerate zone 5, and even those are tricky to overwinter. I lose my rosemary plants too often, and have taken to growing them in large pots and bringing them inside in winter. Inside, they dry out very easily and overwatering becomes a problem inside, so I set the rosemary pot on some rocks and water in a larger container. Water in the lower pot evaporates around the leaves of the rosemary without the soil soaking up the water! Cut, dry and crush the leaves.

Rosemary

Sage (zones 5–8): There are several species of sage; some are in the Artemisia family (like California Sagebrush), but most others are classified in the mint family, Lamiaceae, and can be used like rosemary (which just has recently been added to the Salvia genus/Lamiaceae family!). Most sages are easy to grow, but make sure you know what your species of sage prefers for growing conditions.

Sage

Harvest, dry and crush the leaves; it is the same process for all members of the sage family.

Heather

Chestnut

Hazelnut

Tannins: Tannins are found in the leaves of heather (19% tannins) (zones 3–7) chestnut trees (24.7%) (zones 4–8), hazelnut trees (14.2%) (zones 5–8), oak trees (5.3%) (zones 3–10), and *Sericea lespedeza* (avg. 11%) (zones 5–11) to name a few. Studies across the board indicate feeding *Sericea lespedeza* (SL) hay or high-

Sericea lespedeza

level tannin leaves reduces the parasite load in goats and sheep from 60 percent to 88 percent. However, concentrated feeding (over 75% of the total ration) of high-concentrate SL has resulted in death among cattle in studies. As a guide, then, plants with the highest percents of tannins should never exceed 75 percent of a ration with any livestock. These are great to include in larger pastures where free-feeding on a wide variety of plants is possible. And, with over 90 varieties of oak in North America, these too are easy additions.

Tansy (zones 3–9): I have only had tansy growing as a beneficial plant where fowl can eat it fresh, as needed. It is very easy to grow in many zones, and is a pretty addition to your beneficial plants areas! I cannot offer offer any personal insights to drying/processing this plant.

Tansy

Tarragon (Artemesia family) (zones 4–8): I have not used this plant, but see "Sage," above.

Tarragon

Thyme (Artemisia family) (zones 2–10): With over 150 varieties available, be careful to choose one for your hardiness zone, as well as for your soil. I use a lot of thyme for cooking, even though none of my animals seemed to eat this by choice. But, many varieties grow easily in horrible soil and direct sunlight, and it's an easy addition to pastures and paddocks where animals can nibble when needed.

Thyme

Tobacco (zones 2–10): This highly toxic plant had its greatest success as an antiparasitic with poultry. The USDA had once recommended this plant as a dewormer, but because of its high toxicity, it was removed from veterinary manuals as an internal antiparasitic. However, it's easy to grow in warmer, long-season areas and works well against external pests (lice, mites, fleas, and keds) as a tea-spray. Just soak 4–5 large leaves overnight in a covered container with a gallon of hot water. In the morning, remove the leaves, and use the remaining water as a spray. *Do not use on animals that lick their fur like cats and dogs.* You can also use this to kill pests in the garden.

Turpentine: I own an old veterinary manual from 1919 that lists the doses and administering of turpentine. But, unless it was the apocalypse, I would not use this highly volatile compound.

Tobacco

It is possible to produce your own, but it requires specialized distilling equipment, and the process will not be covered here. Please do further research.

Wormwood (zones 4–8): In the Artemesia family, there are over 300 species, including sage, mugwort, tarragon, and obviously, the many wormwood varieties, which makes it an easy addition to your herb garden, or beneficial plants pasture. Our family cow loved eating just about anything in the Artemisia family, so I had to section-off these plants with electric fencing or she'd eat them to the

Wormwood

ground! She'd even reach over my garden fence to get them, and I had to use cattle panels and electric to keep her from pushing on the garden fence to get at it! I've had the best luck with buying baby wormwood and sage plants, though mugwort was the most easily started from seed. With any wormwords, like most artemesias, cut, dry, and crush the leaves three or four times a season! It's one of the plants I use consistently in dewormers (except *not* with pregnant animals).

Yarrow (zones 3–9): Yarrow can be grown from seed but requires a period of cold to germinate. In the fall or late winter, cast seed directly onto the ground and cover with a light layer of soil and then mulch. Or, in early February, put seeds in a tray containing peat-free compost. The seeds are very small and so should be covered only by a very thin layer of soil. Lightly moisten, then put the tray in

Yarrow

the freezer for a day, and then in the fridge for two to three days. Remove and set in a sunny window. When the seeds have germinated and a small rosette of leaves has formed, transplant individual plants into small pots to grow more before planting outside in late spring. They thrive in full sun and bad soil! Plants can be divided as they fill out and spread. Common yarrow is a weedy species and can become invasive. Wild yarrow grows easily and can also become invasive. My goats seemed to only eat the flower heads, and only a few at a time. A final word on harvesting wild plants - please be cautious of harvesting from roadsides. These areas are laden with heavy metals and other pollutants, as well as herbicides and even pesticides (in urban areas). Many of the plants listed above grow magnificently along the manicured edges of our roadways; Echinacea (purple coneflower), yarrow, bluebonnet, barberry, and many more! But, these should never be harvested and fed to animals due to the danger of contamination.

Performing At-Home Fecal Sample Checks

You may wish to save time and money by performing fecal sample checks at home. You will need to purchase the supplies on the following annotated list, and carefully follow the instructions. It may be a difficult procedure at first, but with practice your technique will improve.

Supplies:

- Microscope with 10X, 40X, and 100X magnification, lighted (I purchased mine used from a homeschooling group and spent around $75)
- Slides and coverslips (can be ordered online)
- Scale that measures in grams (a jewelry or kitchen scale is best)
- Three measuring "beakers" or cups from 25 mL to 50 mL, ideally with a pour-spout
- Timer (many watches and phones already have this option)
- Craft sticks (popsicle sticks)
- Flat-bottomed 15–20 mL test tubes or vials (I use 1.5-oz glass essential oil bottles)
- Small, fine-mesh strainer (mine is from a cookware department and is about 3 inches in diameter with a slight cone-shaped bottom)
- Gauze (I buy the 3–4-inch-wide rolls from a pharmacy)
- Rubber or disposable lightweight gloves
- Reference manual containing pictures of eggs (I use a veterinary parasitology reference manual organized by each breed's specific parasites. You could try to research and/or download pictures online, but this is time consuming and the pictures may not be as clear as those in parasite manuals. Also, while looking under the microscope, it's much quicker to reference a book than to conduct an Internet search. Familiarize yourself with what some of these eggs look like before starting your microscope search.)
- Flotation solution (recipe below)
- Table covering (newspaper or a plastic covering to be used only for this procedure)
- Paper and writing utensil
- Plastic baggies and permanent marker
- Plastic spoons or stainless steel spoons (to be used only for this procedure)

Flotation Solution

Some flotation solutions are made with salt, but I feel they dry out the sample too quickly, so I prefer a sugar solution.

Granulated sugar	454 g (one pound)
Tap water	355 mL (12 ounces, or 1½ cups)

- Dissolve sugar in hot tap water directly, or add sugar to hot water over low heat. After measuring out amount needed, refrigerate the rest to prevent mold growth.

Methodology

1. Collect the sample(s). I take baggies, plastic spoons, rubber gloves, and a permanent marker with me. After donning my gloves, I then follow around the intended animal target, waiting for a fresh sample to fall. You will only need 3 grams (approx. ¾ tsp.). If I'm collecting from more than one animal, I write the name of the animal on the bag with the marker.
2. Back at "the lab" (your table which is covered with protective material), measure out 3 grams (goat, sheep, and rabbit poop will be about 3 to 4 nuggets) and put the poop in a small container (e.g., baby food jar) that can hold 30–50 mL of fluid.
   Note: Before turning on the scale, I set a piece of wax paper or foil on the platform so it zeroes out with the weight of the paper or foil. I then measure the poop sample on the paper/foil to keep my scale clean.
3. With the craft stick, break down the sample well, then add 25 mL of flotation solution. If the sample is really hard, I add a bit of solution as I'm smashing.
4. Stir well, let stand for 2 minutes, then mix well again.
5. Put a 3x3-in. (or 4x4-in.) piece of gauze in the strainer. Set the strainer over a larger glass container and pour the solution into the strainer. You may have to do a little at a time so as not to have overflow.
6. Using a spoon, press the sample against the strainer, forcing out as much liquid as you can.
   Note: I have even made little "bags" out of the gauze by gathering the corners together and then pressing the bag against the strainer, forcing the liquid out. Your gloved hands may get messier using this method.

7. Let this strained solution stand for 2 minutes.

8. Stir thoroughly, then pour the solution into your flat-bottom test tube or vial until the solution is slightly above the vial's top edge, almost to the point of spilling over. If there is not enough liquid to slightly round the top on the vial, add some flotation solution, a little at a time, until it's rounded.

9. Gently place a cover slip on the top of the vial. The solution should spread onto the bottom of the cover slip. If it does not, remove it and add a few more drops of solution.

10. Let this set for 20 minutes. The eggs will float to the top of the "heavy" solution because the eggs are lighter than the sugar-saturated solution.

11. Pull the cover slip straight up off of the vial. Place the cover slip, wet side down, onto the center of the slide.

    Note: If checking more than one animal, mark an edge of the slide with the animal's name using a grease pencil or masking tape.

12. Have your paper/pen ready, next to the microscope. I sometimes have a few "lines" for columns and rows so if I find more than one species of parasite egg, I can write the names down and keep track of the total numbers of each.

13. Place the slide on the microscope platform and set the magnification to 40X.

14. Move the knobs until one of the corners of the cover slip is visible in the eyepiece. Scan the slide, moving it left and right or up and down. When I reach the "end" of one sweep, before I move over to begin the next, I find a "marker" at the far edge of the image to use as a guide (a bubble, hair, etc.).

15. At first, it's difficult to get the hang of moving the slide because it moves opposite of the direction you're turning the knobs. Practice will make it easier.

16. As you move the slide, look for eggs. Use your parasitology guide. You will not see worms. (Remember, the eggs are passed in the poop, and the eggs hatch after hitting the ground.) Mark your paper with "tick" marks for each egg you see, under the species of parasite you think it is.

All animals will have some parasites, and each animal has its own tolerance level. Some can truly tolerate a much higher count than others. But, that said, if the count of any one parasite's eggs is over 20 on one slide, it's best to treat the animal. Remember, that number of eggs you counted was found in only three ounces of poop; think how many are in the whole pile! Even if you recently dewormed an animal, it's okay to dose them again with an antiparasitic blend targeted at the parasite that presented the highest egg count.

Final Thoughts

In summary, these are the most important strategies for you to take away from this book.

1. Rotation, rotation, rotation. Move animals into new pastures and paddocks at least every 60 days. Every 30 days is better, and even more frequently is best during outbreaks or high-load times.
2. Keep stable/coop/cage areas and close-quarter areas free of debris, and clean regularly, preferably after deworming times. Use drying agents like lime and diatomaceous earth.
3. After developing a tailored rotation system around your livestock and land, the next most time-consuming task is keeping pastures and water systems maintained and healthy. Keep water sources clean, and try to keep livestock from drinking from ponds, creeks, puddles, etc. In extremely wet conditions, remember to prohibit livestock from grazing on pastures that have widespread manure and short forage.
4. At least six consecutive days of antiparasitic treatments is best. As a standard deworming practice, it's best to administer antiparasitics for at least three consecutive days (*every* month with housed animals), and follow up with a free-choice lick, with fresh ingredients or herbs, or run livestock through a beneficial plants paddock for an additional three days. Deworming time is greatly reduced by creating bulk mixes and/or boluses, and most ingredients can be purchased online and delivered.

Keep in mind that some animals are more susceptible than others. They may seem to be less vibrant or more sluggish than others in the group. These would be animals for which you'd want to check fecal samples regularly. These are also the ones you would want to consider culling when possible. Usually because of poor eating habits (not avoiding feces-riddled, foul-smelling areas) or because these animals are not consuming antiparasitic forages, they are the ones who will pass along the poor habits to their offspring. Remember this as you improve your herd or flock.

With proper rotation and monthly deworming, fecal tests will become fewer and fewer and thwarting internal parasites will no longer be a full-time job.

Pictured from left to right: The author's Jersey dairy cow, Rosie; her favorite sow, Lizzie, with a litter of hours-old piglets; and Arrown, her lead dairy doe. Healthy livestock should be shiny, free of manure, and content.

CHART D

Common Name(s)	Scientific Classification
bankrupt worm	(C), O-Rhabditida, F-Trichostrongylidae, G-Trichostrongylus
barber pole worm, wireworm, red stomach worm	(C), O-Rhabditida, F-Haemonchidae, G-Haemonchus
bloodworm (red worm), large strongyles	(C), O-Rhabditada, F-Strongyloidae, G-Strongylus
bots	Ph-Arthropoda, Cl-Insecta, O-Diptera, F-Oestridae, G-Gasterophilus
brown stomach worm	(C), O-Rhabditada, F-Trichostrongyladae, G-Ostertagia
caecal worm	(C), O-Ascaridida, F-Heterakidae, G-Heterakis
coccidia (Cryptosporidium and Eimeria)	K-Chromalveolate, Ph-Apicomplexa,Cl-Coccidia, O-Eucoccidiorida,F-diff.for Crypto & Eimeria
giardia	K-Excavata, Ph-Metamonada, O-Diplomonadida, F-Hesamitidae, G-Giardia
gapeworm	(C), O-Rhabditida, F-Syngamidae, G-Syngamus
gullet worm	(C), O-Spirunia, F-Gongylomematidae, G-Gongylonema
hookworm	(C), O-Rhabditida, F-Ancylostomatidae, G-Bunostomum
liver fluke	Ph-Platyhelmenthes, Cl-Trematoda, O-Plagiorchiida, F-Fasciolidae, G-Fasciola
lungworm	(C), O-Rhabditida, F-Dityocaulidae, G-Dictyocaulus
nodular worm	(C), O-Rhabditida, F-Cloacinidae, G-Oesophagostomum
pinworm	(C), O-Oxyurida, F-Oxyuridae, G-Skrjabinema
roundworm (large)	(C), O-Ascaridida, F-Ascarididae G-Ascaris
small intestine worm	(C), O-Rhabditida, F-Cooperiidae, G-Cooperia
small red worm (small strongyles)	(C), O-Spirurida, F-Habronematidae, G-Habronema
stomach hair worm	(G), Gordea, F-Gordiidae, G-Gordius
tapeworm	Ph-Platyhelmenthes, Cl-Cestoda, O-Cyclophyllidae, F-Taeniidae, G-Echinococus
threadworm	(C), O-Rhabdita, F-Strongyloididae, G-Strongyloides
trichinosis	(E), O-Trichocephalida, F-Trichinellidae, G-Trichinella
whipworm	(E), O-Trichocephalida, F-Trichuridae G-Trichuris

Kingdom = K; Most are *Animalia* unless noted, **Phylum = Ph**; Most are *Nematoda* unless noted, **Class**=Cl and, if Nematoda then class is (C)=Chromadorea (E)=Enoplea (G)=Gordioida, **Order**=O, **Family**=F, **Genus**=G, S=**Species** are not listed since most are host-specific, and up to several hundred species can exist in one genus

Animals Affected	Area Affected	Treatment (suggestion from research) See individual ingredients list for more
C,G,S	small intestine	fineleaf fumitory, white wormwood
C,G,S	abomasum	white wormwood (*A. herba alba*), curcurbits, COWP, orange oil, carrot seed, fennel
H	large intestine, arteries	fineleaf fumitory, clove, wormwood
H	stomach, intestines, sinuses	diatomaceous earth, garlic, wormwood
C,G,S	abomasum	COWP for goats, tannins, fineleaf fumitory
Po	caecum	aloe, pineapple, tobacco
All, esp. young	intestine	aloe, charcoal, oregano, mustard seed, tannins, terpines, rosemary, sage, thyme
All, esp. young	intestine	berberines, garlic, grapefruit seed extract, tannins
Po	trachea	orange oil, grapefruit seed extract
Po	mouth, esophagus	grapefruit seed extract
C,G,S, Cat, Dog	small intestine	wormwood (*A. vulgaris*)
C,G,S	liver, gall bladder, bile ducts	conifers, curcurbits, ginger
C,G,H, S	lungs	wormwood (*A. vulgaris*), carrot seed, garlic
C,G,P,S	small and large intestine	fineleaf fumitory, wormwood
H.R	large intestine	curcurbits, black walnut hulls
H, Po, Cat, Dog	small intestine	garlic, ginger, curcurbits, black walnut hulls, fineleaf fumitory, tansy
C,G,S	small intestines	fineleaf fumitory, tannins, COWP
H	stomach, lungs (migratory)	wormwood, carrot seed, garlic
C,G,H,S	abomasum	fineleaf fumitory, wormwood
All	intestine	curcurbits, fern (male), black walnut hulls
C,G,H, Po, S	small intestine	ginger, clove, wormwood (*A. vulgaris*), fennel, garlic
P	muscles, intestines	wormwood
Po, Cat, Dog	large intestine	fineleaf fumitory, wormwood

C = Cattle, G = Goat, H = Horse, Po = Poultry (chickens, ducks, geese, turkeys), P = Pig, S = Sheep, R = Rabbit

COWP=Copper Oxide Wire Particles (Copper Bolus)

Resources

1. *The Association of Natural Biocontrol Producers* (ANBP). www.anbp.org
2. *ATTRA Sustainable Agriculture Program,* developed and managed by the National Center for Appropriate Technology. attra.ncat.org One particularly good article can be found at https://attra.ncat.org/what_can_you_tell_me_about_organic_paras/
3. *Sustainable Agriculture Research and Education* (SARE). www.sare.org Especially check out these headings on their website's drop-down menu: "Project Reports," "Learning Center," and "About SARE/Professional Development."
4. *The Complete Herbal Handbook for Farm and Stable,* fourth ed. (1991). Juliette de Baïracli Levy
5. *"PARA-Site,"* an interactive multimedia electronic resource developed by the Australian Society of Parasitology. http://parasite.org.au/para-site/introduction/—Besides just having *lots* of information, I like their description of host-spec-ificity. (On the left side of the home page, click on "Introduction," then click on "Overview of Parisitology" at the bottom of that page.)
6. *The National Center for Biotechnology Information* (NCBI). https://www.ncbi.nlm.nih.gov/ There is especially interesting information on scientific classification (click on the "Taxonomy" link on the left side of the page).
7. *Planet Natural.* www.planetnatural.com Beneficial mini-predatory insects for sale.
8. The References (page 81). Google the name of the researcher, the title, and the date. Many of the articles and books will lead to even more interesting and useful references.

References

Akhtar M. S., and I. Ahmad. "Comparative efficacy of *Mallotus philippinensis* fruit (Kamala) or Nilzan® drug against gastrointestinal cestodes in Beetal goats." *Small Ruminant Research* (1992): 121–128.

Al-Shaibani, I. R. M., et al. "Anthelmintic activity of *Fumaria parviflora* (Fumariaceae) against gastrointestinal nematodes of sheep." *International Journal of Agriculture and Biology* 11 (2009): 431–436.

Amin, M. R., et al. "In vitro anthelmintic efficacy of some indigenous medicinal plants against gastrointestinal nematodes of cattle." *Journal of the Bangladesh Agricultural University* 7 (1) (2009): 57–61.

Anderson, David, DVM, MS, DACVS. "Information For Parelaphostrongylus Tenuis (Meningeal Worm)" Department of Large Animal Clinical Sciences, College of Veterinary Medicine, University of Tennessee, 2013.

Anderson, N., et al. Impact of gastrointestinal parasitism on pasture utilisation by grazing sheep. Melbourne: Australian Wool Corporation/CSIRO Technical Publication, 1987.

Barbercheck, Mary. "Insect-Parasitic Nematodes for the Management of Soil-Dwelling Insects." Penn State College of Agricultural Sciences research, extension, 2015.

Barrell, G. K. (ed.) "Sustainable control of internal parasites in ruminants." Animal Industries Workshop. Lincoln University, New Zealand, 1997.

Bastidas, G. J. "Effect of ingested garlic on necator americanus and ancylostoma caninum." *American Journal of Tropical Medicine and Hygiene* 18 (6) (1969): 920–923.

"Biological control of gastro-intestinal nematodes of ruminants using predacious fungi." Danish Centre for Experimental Parasitology & Food and Agriculture Organization of the United Nations, Workshop proceedings, 1997.

Cabaret, J., et al. (2002) "Managing helminths of ruminants in organic farming." *Veterinary Research* 33 (5) (2002): 625–640.

Caner, A., et al. "Comparison of the effects of *Artemisia vulgaris* and *Artemisia absinthium* growing in western Anatolia against trichinellosis (*Trichinella spiralis*) in rats." *Experimental Parasitology* 119 (1) (2008):173–9.

Chandrakesan, P., et al. "Efficacy of a herbal complex against caecal coccidiosis in broiler chickens." Department of Veterinary Parasitology, Veterinary College and Research Institute, Namakkal, Tamil Nadu, India; *Veterinarski Arhiv* 79 (2) (2009): 199–203.

Chopra, R. N., et al. *Glossary of Indian medicinal plants.* New Delhi: Council of Scientific & Industrial Research, 1956.

Coffey, Linda, et al. Tools for Managing Internal Parasites in Small Ruminants: Sericea Lespedeza. NCAT/ATTRA and Southern Consortium for Small Ruminant Parasite Control, IP316, Slot 315, Version 112007, 2007.

"Control of External Parasites of Sheep and Goats" Ethiopia Sheep and Goat Productivity Improvement Program (ESGPIP) Technical Bulletin 41, 2010.

Cormanes, Joan Marie, et al. "In vivo anthelmintic activity of pineapple (*Ananas comosus Merr.*) fruit peeling juice in semi-scavenging Philippine native chicken naturally co-infected with *Ascaridia galli* and *Heterakis gallinarum*." *Livestock Research for Rural Development* 28 (5) (2016).

Coleby, Pat. *Natural Goat Care.* Texas: Acres USA Publishers.

De, S., and P. K. Sanyal. "Biological Control of Helminth Parasites by Predatory Fungi." Department of Parasitology, College of Veterinary Science & Animal Husbandry, Indira Gandhi Agricultural University. *VetScan* 4 (1) (2009), Article 31.

Deore, P. A. "Different approaches for alternate systems of animal health care." In Proceedings of an International Seminar on an Integrated Approach for Animal Health, Calicut, India, 1999.

Duke, J. A. *CRC Handbook of Medicinal Herbs.* Boca Raton, Florida: CRC Press, 1985.

Dung Beetle Fact Sheet. Dung Beetles Direct, University of Bristol, and National Environment Research Council. www.dungbeetlesdirect.com

Durrani, F. R., et al. "Using Aqueous Extract of Aloe Gel as Anticoccidial and Immunostimulant Agent in Broiler Production." *Sarhad Journal of Agriculture* 24 (4) (2008).

Duval, Jean. "The control of internal parasites in ruminants." McGill University Ecological Agricultural Projects, 1994.

Ethnoveterinary medicine in Asia: An information kit on traditional animal health care practices. 4 vols. International Institute of Rural Reconstruction, Silang, Cavite, Philippines, 1994.

Ethnoveterinary medicine in Kenya. A field manual of traditional animal health care practices. Nairobi: Intermediate Technology Kenya and the International Institute of Rural Reconstruction, 1996.

Ferris, H., et al. "Plant Sources of Chinese Herbal Remedies: Effects on *Pratylenchus vulnus* and *Meloidogyne javanica*." *Journal of Nematology* 31 (3) (1999).

Fielding, D. "Validation of ethnoveterinary medicine." In Proceedings of an International Seminar on an Integrated Approach for Animal Health, Calicut, India, 1999.

"Fungi as a Biological Control in Grazing Livestock." *Professional Animal Scientist* 21 (1) (Feb. 2005): 30–37.

Gamble, H. Ray. "Trichinae: Pork Facts—Food Quality and Safety." USDA, Agricultural Research Service, Parasite Biology and Epidemiology Laboratory, 1998.

Gandolf, R., and J. Ter Beest. "Meningeal Worm (*Parelaphostrongylus tenuis*), *Ageratum conyzoides* (aster?), *Annona muricata*, *Ocimum gratissimum* (eugenol/basil)." Fact Sheet: American Association of Zoo Veterinarians Infectious Disease Committee, 2011.

"Garlic as an anthelmintic." *Vet Record* 65 (28) (1993): 436.

Geiger, Flavia, et al. "Insect abundance in cow dung pats of different farming systems." *Diptera Entomologische Berichten* 70 (4) (2010): 106–110.

Grieve, M. *A Modern Herbal.* Volume 1. New York: Dover, 1971.

Hammond, J. A., et al. (1997). "Prospects for plant anthelmintics in Tropical Veterinary Medicine." *Veterinary Research Communication* 21 (1997): 213–228.

Idris, U. E., S. E. I. Adam, and G. Tartour. "The anthelmintic efficacy of Artemisia herba-alba against Haemonchus contortus infection in goats." *National Institute of Animal Health Quarterly* 22 (3) (1981): 138–143.

Jacob, Jacquie, and Tony Pescatore. "Natural remedies for poultry diseases common in 'natural' and 'organic' flocks." Extension Paper: Cooperative Extension Service, College of Agriculture, University of Kentucky, 2011.

Jabbara, A., et al. "Anthelmintic activity of *Chenopodium album* (L.) *and Caesalpinia crista* (L.) against trichostrongylid nematodes of sheep." *Journal of Ethnopharmacology* 114 (1) (November 2007): 86–91.

Julien, J., et al. "Extracts of the ivy plant, *Hedera helix*, and their anthelmintic activity on liver flukes." *Planta Medica* 51 (1985): 205–208.

Kambewa, B. M. D. "Integration of Indigenous Veterinary Remedies in Farming Systems in Malawi." Proceedings of the International Seminar, Kozhikode, 1999.

Kaufman, P. E., et al. "External Parasites on Beef Cattle." Document ENY-274; Department of Entomology and Nematology, UF/IFAS Extension. Original publication date May 1995. Revised April 2011 and April 2017.

Korinek, C. J., DVM. *The Veterinarian*, 4th Edition. Toronto, Canada: The Gerlach-Barklow Co., 1917.

Lans, C. et al. "Ethnoveterinary medicines used for horses in Trinidad and British Columbia, Canada." *Journal of Ethnobiology and Ethnomedicine* 2 (2006): 31.

Lans, Cheryl, et al. (2007) "Ethnoveterinary medicines used to treat endoparasites and stomach problems in pigs and pets in British Columbia, Canada." *Journal of Ethnobiology and Ethnomedicine* (Oct. 2007): 148 (3–4): pp 325–340.

Larsen, M. "Biological control of helminthes." *International Journal for Parasitology* 29 (1999): 139–146.

"List of Medicinal Herbs Used for Antihelminthics" (2010) *International Journal of Pharmaceutical Sciences Review and Research* 5 (3) (2010): Article 007.

Maday, John. "Control External Parasites, Prevent Disease." *Bovine Veterinarian,* (Mar. 2018), online article: https://www.bovinevetonline.com/article/control-external-parasites-prevent-disease.

Marin Municipal Water District Vegetation Management Plan (DRAFT-1/1/2010). Glyphosate: Herbicide Risk Assessment 3–1 Corte Madera, CA (Jan. 2010).

Macleod, George. *A Veterinary Materia Medica and Clinical Repertory: With a Materia Medica of the Nosodes.* London: Random House UK, 1997.

Mbaria, J. M., et al. "Acute toxicity of Pyrethrines in Red Masai sheep and New Zealand White rabbits." *Bulletin of Animal Health and Production in Africa* 42 (1994): 217–221.

Modini, Laura, et al. "Longevity of Cryptosporidium oocysts in fresh and sea water at environmental temperatures." 2nd International Conference on Parasitology, Manchester, UK, 2016.

Morales, M. R. "Improving Small Ruminant Grazing Practices." Research Update and Workshop Proceedings; Medicinal Botanicals Program, Mountain State University, Beaver, West Virginia, 2009.

Mostofa, M., et al. "Epidemiology of gastrointestinal helminth parasites in small ruminants in Bangladesh." Sustainable Parasite Control in Small Ruminants. Proceedings of a workshop, Bogor, Indonesia, 22–25 April 1996. ACIAR Proceedings 74: 105.

Mostafalou, Sara, and Mohammad Abdollahi. "Pesticides and human chronic diseases: Evidences, mechanisms, and perspectives." *Toxicology and Applied Pharmacology* 268 (2013): 157–177.

Mwale, M., et al. "Effect of Medicinal Plants on Hematology and Serum Biochemical Parameters of Village Chickens Naturally Infected with *Heterakis gallinarum*." *Bangladesh Journal of Veterinary Medicine* 12 (2) (2014): 99–106.

Nadkarni, A. K. *Indian Materia Medica*. 3rd Edition. Bombay, India: Popular Prakashan, 1954.

Ndi, A., et al. "The anthelmintic efficacy of some indigenous plants in the Northwest Province of Cameroon." *Revue d Elevage et de Medicine Veterinaire des Pays Tropicaux* 52 (2) (1999): 103–106.

Niezen, J. H., et al. "Growth and gastrointestinal nematode parasitism in lambs grazing either lucerne (*Medicago sativa*) or sulla (*Hedysarum coronarium*) which contains condensed tannins" *Journal of Agricutural Science* 125 (2): 281–289.

Papchenkov, N. Y. "*Tanacetum vulgare* seed and naphtaman against Nematodirus infections in sheep." *Veterinarya Moskva* 45 (8) (1968): 48–49.

Piyush, Jain, et al. "Anthelmintic Potential of Herbal Drugs." *International Journal of Research and Development in Pharmacy and Life Sciences* 2 (3) (2013): 412–427.

Rahmann, G., and H. Seip. "Bioactive forage and phytotherapy to cure and control endo-parasite diseases in sheep and goat farming systems—a review of current scientific knowledge." Institute of Organic Farming, Agricultural Research Centre.

Landbauforschung Völkenrode (2007): 285–295, Bundesforschungsanstalt für Landwirtschaft, Westerau, Germany. (Translated version available on academia.edu.)

Robertson, L. J., et al. "Survival of *Cryptosporidium parvum* Oocysts under Various Environmental Pressures." *Applied Environmental Microbiology* 58 (11) (1992): 3494–3500.

Sanyal, P. K. "Biological control by predacious fungi for integrated parasite management in livestock." In Proceedings of an International Seminar on an Integrated Approach for Animal Health, Calicut, India, 1999.

Satrija, F., et al. "Effect of Papaya latex against *Ascaris suum* in naturally infected pigs." *Journal of Helminthology* 68 (1994): 343–346.

Satyavati, G. V., et al. "Medicinal plants of India- Vol 2." Indian Council of Medicinal Research, New Delhi, 1987.

Seddiek, S.A., et al. "Anthelmintic activity of the white wormwood, *Artemisia herba-alba*, against *Heterakis gallinarum* infecting turkey poults." *Journal of Medicinal Plants Research* 5 (16) (2011): 3946–3957.

Shalaby, H. A., et al. "The role of dogs in transmission of *Ascaris lumbricoides* for humans." *Parasitology Research* 106 (5) (2010): 1021–1026.

Sharma, L. D., et al. "In vitro anthelmintic screening of indigenous plants against *Haemonchus contortus* of sheep and goats." *Indian Journal of Animal Research* 5 (1) (1971): 33–38.

Silva, Manoel Eduardo, et al. "Predatory activity of *Butlerius* nematodes and nematophagous fungi against *Haemonchus contortus* infective larvae." *Revista Brasileira Parasitologia Veterin*ária 26 (1) (2017).

Soder, K. J. and L. A. Holdon. Use of Nematode-Trapping. *Professional Animal Scientist* 21 (1) (2005): 30–37.

Spencer, Robert. "Using Dung Beetles to Maintain a Healthy Pasture Ecosystem." UNP-0137, Alabama A&M University Extension, 2013.

Squires, Jill M., et al. "Efficacy of an orange oil emulsion as an anthelmintic against *Haemonchus contortus* in gerbils (*Meriones unguiculatus*) and in sheep." *Veterinary Parasitology* 172 (2010): 95–99.

Swiger, Sonja L. Managing External Parasites of Texas Cattle; AgriLIFE Extension, Texas A & M System; E-570 (August 2012).

Tariq, K. A., et al. "Anthelmintic efficacy of *Achillea millifolium* (yarrow) against gastrointestinal nematodes of sheep: in vitro and in vivo studies." Indian Institute of Integrative Medicine, Council of Scientific and Industrial Research, 2008.

Tariq, K. A., et al. "Anthelmintic activity of extracts of *Artemisia absinthium* against ovine nematodes." *Veterinary Parasitology* 160 (1 & 2) (2009): 83–88.

Thamsborg, Stig Milan, et al. "Alternative approaches to control of parasites in livestock: Nordic and Baltic perspectives." Danish Centre for Experimental Parasitology, Department of Veterinary Disease Biology, Faculty of Life Sciences, University of Copenhagen, Denmark, 52 (1) (Oct. 2010).

Thomas, Michelle L. "Dung Beetle Benefits in the Pasture Ecosystem." ATTRA/ NCAT F (www.attra.org/attra-pub/PDF/dungbeetle.pdf). (Accessed July 15, 2020.)

Veerakumari, L. "Botanical Anthelmintics." *Asian Journal of Science and Technology* 6 (10) (2015): 1881–1894.

Waller, P. J., and Margaret Faedo. "The prospects for biological control of the free-living stages of nematode parasites of livestock." *Journal of Parasitology* 26 (8/9) (1996): 915–925.

Waller, P. J., et al. "Plants as De-Worming Agents of Livestock in the Nordic Countries: Historical Perspective, Popular Beliefs and Prospects for the Future." *Acta Veterinaria Scandinavica* 42 (1) (2001): 31–44.

Wang, G., and M. P. Doyle. "Survival of Enterohemorrhagic *Escherichia coli* O157:H7 in Water." *Journal of Food Protection* 61 (6) (1998): 662–667.

Wells, Ann. "Integrated Pest Management for Livestock." Appropriate Technology Transfer for Rural Areaa (ATTRA), Livestock Systems Guide (April 1999).

Wells, Ann., et al. "Sustainable sheep production." Appropriate Technology Transfer for Rural Areas (ATTRA), Livestock Systems Guide. (May 2000).

Whitlock. J. H., et al. "The relationship of diet to the development of *Haemonchosis* in sheep." *Journal of the American Veterinary Medical Association* (January 1943): 34–35.

Xiao, Lihua, et al. "Cryptosporidium Taxonomy: Recent Advances and Implications for Public Health." *Clinical Microbiology Review* 17 (1) (2004): 72–97.

Index

A

acorn, 36

acorn squash, 36

alkaloids, 36

aloe, 39, 55

Anaplasma marginale, 11–13

angelica, 35

anise, 22, 35

antiparasitics, 34–53

aqueous liquid extract (ALE), 47

ascaris, 40, 43, 44

atlas squash, 36

B

banana squash, 36

bankrupt worm, 10, 78

barber pole worm, 6, 23–24, 40, 41, 42–43, 44–45, 78

barberry, 36, 39, 55

berberine, 36, 39, 55

birdsfoot trefoil, 36

black walnut, 36, 40, 56

blind staggers, 13–14

bloodworm, 78

bolus, 47

bots, 78

brown stomach worm, 6, 78

butternut squash, 36

C

cabbage, 57

caecal worm, 78

carrot seed, 35, 40, 57–58

carvacrol, 22

cats, 10, 48

cattle, 10, 48

cayenne, 40, 58

cheese squash, 36

chicken

    combs, 5, vii

    waddles, vii

chrysanthemum, 40, 58

cinnamon, 22

Classification of Organisms, 2–4

clove, 22, 40, 59

coccidiosis, 5, 39, 46, 78

cocozelle, 36

collards, 35, 40, 57

conifers, 36, 41, 59

copper, 41, 59

coriander, 35

coumarin, 35

crookneck squash, 36

cucurbitin, 35, 36, 41, 59–60

cumin, 35

cushaw, 36

D

diatomaceous earth (DE), 16–17, 18, 33

dill, 35

dipping, 15–21

dock, 36

dogs, 10, 48
dosages, 47–48
dung beetles, 28
dusting powder, 20

E

echinacea, 41, 54, 60
enzyme spray, 18
estragole, 22
eugenol, 22
eukaryotic cells, 2–3, 21–22
external parasites, 11–22
eyelids, viii

F

fecal sample checks, 73–75
fenchone, 35
fencing, 31–32
fennel, 35, 42, 60–61
fern, 42, 61
fineleaf fumitory, 42–43, 62
flotation solution, 74
flukes, 6
fly-repellent spray, 19–20
fowl, 10, 48
fungi, 29

G

gapeworm, 78
garlic, 22, 43, 63
garlic powder, 18
geranium, 22
giardia, 39, 43, 44–45, 78
ginger, 43, 63
goat
    common parasites of, 10

eyelids, viii
    ingredient combinations for, 48–49
goldenseal, 36, 39, 55
goosefoot, 43, 64
gourds, 36
grapefruit seed, 43
grass, 27
grazing habits, 24–25
gullet worm, 78

H

heartworm, 43
hookworm, 6, 78
horn flies, 11
horseradish, 40
horses, 10, 41, 49
host-transference, 3–4

I

Indian ginger, 43, 64

J

juglone, 40
juniper berries, 43

K

kale, 35, 40, 57

L

lactones, 35
life cycles, 4–10
lime, 33
limonene, 35
liver fluke, 7, 41, 45, 78
lungworm, 6, 40, 43, 78
lupines, 36, 44, 65

M

mace, 22
mallow, 36
manure, 28–29
meningeal worms, 6–7, 31, 39
molasses, 47
mugwort, 46
mustards, 35, 40, 57

N

nasturtium, 35, 44, 57, 65–66
neem oil, 18–19
nematodes, 6, 28
nodular worm, 6, 42–43, 78
nose botfly, 13–14
nutmeg, 22

O

oak, 36
oils, 18
onion, 22
orange oil extract, 44, 66
oregano, 22, 44, 66
Oregon grape, 36, 39, 55

P

parasite(s)
in Classification of Organisms, 2–4
external, 11–22
history of term, 1
life cycles of, 4–10
parasitology, 1
parsley, 35
parsnip, 35
pasture management, 23–32
pasture rotation, 23–24, 26, 32–33

pigs, 10, 26, 30, 49
pigweed, 43, 64
pinworm, 40, 41, 78
prokaryotic cells, 2–3, 21–22
protozoa, 4
pumpkin, 36
pyrethrum, 19, 40, 58

R

rabbits, 10, 49
radish, 35, 40
Redi, Francesco, 1
red stomach worm, 78
rosemary, 44, 67
roundworm, 40, 41, 42–43, 78
rutabaga, 35, 40

S

sage, 44, 67
santonin, 35
scabies, 14–16
sesquiterpene, 35
sheep
common parasites of, 10
eyelids, viii
ingredient combinations for, 48–49
small intestine worm, 6, 78
small red worm, 78
snakeroot, 43, 64
soap, 18
squash, 36, 41, 59–60
squash seeds, 35
Steinernema carpocapsae, 28
stomach hair worm, 6
strongyles, 40, 44, 45
sulfur, 16, 22, 33, 35

summer pasture, 25–26

sweet Annie, 46

T

tannins, 22, 36, 44–45, 68–69

tansy, 35, 40, 45, 69

tapeworms, 16, 40, 41, 42, 78

tarragon, 35, 45, 70

terpene, 36, 41

terpenoids, 35

threadworm, 6, 8, 9, 24, 42, 78

thujone, 35

thyme, 35, 45, 70

thymol, 22, 44

tobacco, 45, 70–71

traps, 20–21

Trichinella spiralis, 9–10

trichinosis, 9, 78

turnip, 35, 40

turpentine, 41, 45, 71

V

van Beneden, Pierre-Joseph, 1

van Leeuwenhoek, Antonie, 1

Virginia mammoth, 36

W

watercress, 44

water management, 31

whipworm, 42–43, 78

wireworm, 78

worm, as term, 3

wormwoods, 35, 45, 46, 71–72

Y

yarrow, 46, 54, 72

yellow root, 36

Z

zucchini, 36

Notes

Notes

Notes

Notes

Notes

Notes

Notes

Notes

Notes

Notes

Notes

Notes